Research on Electrocatalytic Materials for Hydrogen Evolution and Oxygen Evolution

Research on Electrocatalytic Materials for Hydrogen Evolution and Oxygen Evolution

Guest Editors

Jiangtian Li
Deryn Chu

MDPI

Basel • Beijing • Wuhan • Barcelona • Belgrade • Novi Sad • Cluj • Manchester

Guest Editors
Jiangtian Li
United States Army Combat Capabilities Development Command Army Research Laboratory
Aberdeen
USA

Deryn Chu
United States Army Research Laboratory-Sensors and Electron Devices Directorate
Adelphi
USA

Editorial Office
MDPI AG
Grosspeteranlage 5
4052 Basel, Switzerland

This is a reprint of the Special Issue, published open access by the journal *Materials* (ISSN 1996-1944), freely accessible at: https://www.mdpi.com/journal/materials/special_issues/I3GQ4V8VIZ.

For citation purposes, cite each article independently as indicated on the article page online and as indicated below:

Lastname, A.A.; Lastname, B.B. Article Title. *Journal Name* **Year**, *Volume Number*, Page Range.

ISBN 978-3-7258-7575-7 (Hbk)
ISBN 978-3-7258-7576-4 (PDF)
https://doi.org/10.3390/books978-3-7258-7576-4

Contents

materials

MDPI

Editorial

Research on Electrocatalytic Materials for Hydrogen Evolution and Oxygen Evolution

Jiangtian Li * and Deryn Chu *

U.S. Army DEVCOM Army Research Laboratory, Adelphi, MD 20783, USA
* Correspondence: jiangtian.li.ctr@army.mil (J.L.); deryn.d.chu.civ@army.mil (D.C.)

There is growing interest in pursuing more sustainable energy sources to replace traditional fossil fuels. Hydrogen stands out as a promising option because it can store and release energy efficiently and produces only water when used as fuel [1]. Conventional hydrogen production primarily relies on natural gas reformation, a process that demands substantial investment in infrastructure and transportation. As an alternative, water electrolysis using electricity offers a more adaptable alternative. This method splits water into hydrogen and oxygen, producing high-purity hydrogen suitable for applications such as fuel cells, where additional purification is unnecessary [2,3]. Furthermore, when driven by indigenous energy sources such as solar, wind, or hydropower, water electrolysis can strengthen energy resilience and provide capabilities that reduce—and potentially eliminate—reliance on fossil fuel supply. However, the inherent kinetic barriers to hydrogen evolution reactions (HERs) and oxygen evolution reactions (OERs) demand highly efficient and cost-effective electrocatalysts to make them viable at scale. To overcome this, researchers are working to develop advanced materials that can speed up these reactions efficiently and affordably, but this remains a major challenge in the field [1–4]. This Special Issue, 'Research on Electrocatalytic Materials for Hydrogen Evolution and Oxygen Evolution', gathers ten original research papers that highlight recent advances in catalyst design, synthesis, and evaluation for the HERs and OERs used in water splitting. The contributions span a range of strategies, from atomic-level surface modification and heterostructure engineering to waste valorization and electrolyte design, offering insights into our fundamental understanding of this topic, as well as practical implementations.

For large-scale applications, the cost of catalysts must be considered alongside their catalytic activity. Currently, state-of-the-art water-splitting catalysts still heavily rely on noble metals. Therefore, the development of efficient, non-precious metal catalysts is in high demand. In this Special Issue, Leiva-Guajardo et al. demonstrate the use of industrial copper slag waste as a HER electrode in saline water electrolysis. Its porous structure facilitates ion transport and enhances HER kinetics while offering environmental remediation through waste reuse. Bojîncă et al. compared Ni catalysts produced via pulsed current deposition from Watts and citrate baths. Ni deposited from citrate exhibits finer grains and superior HER performance, highlighting electrolyte control as a key synthesis lever. In cases where noble metals are used, developing strategies to minimize their loading is essential for reducing costs. Single-atom catalysts (SACs) have emerged as a promising solution, offering advantages such as uniform active sites, high atom utilization, and excellent catalytic activity. Wu et al. designed a sequential underpotential deposition approach to anchor Cu and Pt single atoms onto N-doped graphene, achieving high turnover frequencies in alkaline HERs due to maximized active site exposure. Zhang et al. construct Mo_2C nanoparticles encapsulated in MoS_2 layers to form synergistic interfaces, enhancing charge transport and hydrogen adsorption, resulting in robust acidic HER performance. To investigate the underlying mechanism, Li, Chu et al. employed XPS depth profiling, revealing that P-doping facilitates the reconstruction of sulfide catalysts into Ni-O species during alkaline HERs. This transformation boosts catalytic activity by modulating the electronic structure and *d*-band center. In addition to catalyst improvement, Perović et al. introduced Ni, Zn, and Mo ions into an alkaline electrolyte as ionic activators, resulting in enhanced HER activity. This in situ activation method reduces overpotentials and energy costs, showcasing a potential low-cost improvement route.

Regarding the alkaline oxygen evolution half-reaction, transition-metal-based layered double-hydroxides (LDH) represent one of the most active classes of OER catalysts. In this Special Issue, Li et al. investigate the effect of Fe, Mn, Cu, and Zn doping on NiCo-layered double-hydroxides. Fe-doped catalysts exhibit the best OER performance due to their improved conductivity and nanosheet dispersion. This study offers significant contributions to the development of efficient electrocatalysts for OERs, advancing our understanding of key design principles for enhanced catalytic performance. Yang et al. fabricated composite catalysts using a short plasma process, forming a defect-rich interface between NiFe-LDH and PbO_2 on reduced graphene oxide. This structure exhibits excellent OER activity with low overpotentials. Although an OER is thermodynamically more favorable in alkaline electrolytes, it remains a significant scientific challenge in acidic environments, especially when paired with a Pt cathode for acidic water splitting, due to issues related primarily to catalyst stability. Kuang et al. presented a CeO_2-Ir hybrid catalyst with strong interfacial coupling on carbon nanotubes. This structure achieves high OER activity and durability in acidic media, using significantly reduced Ir content. Lastly, Sun et al. developed porous $NiTiO_3$-BiOBr composites that exhibit efficient photocatalytic activity under visible light. A type-II heterojunction facilitates charge separation and extends photocatalytic lifetime.

The ten contributions showcased in this Special Issue illustrate the breadth and depth of current research efforts in the field of electrocatalyst development, from atomic-level design to large-scale waste reuse. They offer valuable insights into the structure–activity relationships, reaction mechanisms, and performance engineering concerns present in diverse electrochemical conditions. We thank all of the contributing authors, peer-reviewers, and the *Materials* editorial team for their dedication and support. We hope that this Special Issue will serve as a vital reference for researchers working to realize clean and efficient hydrogen technologies.

Conflicts of Interest: The authors declare no conflict of interest.

References

1. Yue, M.; Lambert, H.; Pahon, E.; Roche, R.; Jemei, S.; Hissel, D. Hydrogen energy systems: A critical review of technologies, applications, trends and challenges. *Renew. Sustain. Energy Rev.* **2021**, *146*, 111180. [CrossRef]
2. Service, R.F. The parting of water. *Science* **2025**, *387*, 354–357. [CrossRef] [PubMed]
3. Imran, S.; Hussain, M. Emerging trends in water splitting innovations for solar hydrogen production: Analysis, comparison, and economical insights. *Int. J. Hydrogen Energy* **2024**, *77*, 975–996. [CrossRef]
4. Perumal, S.; Pokhrel, I.; Muhammad, U.; Shao, X.; Han, Y.; Kim, M.; Lee, H. Recent advances in electrochemical water splitting electrocatalysts: Categorization by parameters and catalyst types. *ACS Mater. Lett.* **2024**, *6*, 3625–3666. [CrossRef]

materials

MDPI

Article

Porous Rod-like $NiTiO_3$-BiOBr Heterojunctions with Highly Improved Visible-Light Photocatalytic Performance

Kaiyue Sun, Mengchao Li, Hualei Zhou *, Xiaohui Ma and Wenjun Li

Beijing Key Laboratory for Science and Application of Functional Molecular and Crystalline Materials, Department of Chemistry and Chemical Engineering, University of Science and Technology Beijing, Beijing 100083, China; sunkaiyue12345@163.com (K.S.); limengchao2020ks@163.com (M.L.); 18813010169@163.com (X.M.); wjli_ustb@163.com (W.L.)

* Correspondence: hlzhou@ustb.edu.cn

Abstract: $NiTiO_3$-BiOBr heterostructured photocatalysts were constructed via precipitation, calcination and hydrothermal treatments. Various characterizations demonstrated that BiOBr nanosheets were decorated on $NiTiO_3$ nanoparticals, forming porous rod-like heterojunctions. Compared with independent $NiTiO_3$ and BiOBr, the composites with optimal BiOBr content presented highly improved visible-light photocatalytic efficiency. The degradation rates of Rhodamine B (RhB) and tetracycline (TC) reached 96.6% in 1.5 h (100% in 2 h) and 73.5% in 3 h, which are 6.61 and 1.53 times those of $NiTiO_3$, respectively. The result is an improved photocatalytic behavior from the formation of heterojunctions with a large interface area, which significantly promoted the separation of photogenerated carriers and strengthened the visible-light absorption. Based on the free radical capture experiments and band position analysis, the photodegradation mechanism of type-II heterojunction was deduced. This study provides a new way to fabricate highly efficient $NiTiO_3$-based photocatalysts for degrading certain organics.

Keywords: photocatalysis; heterojunction; porous heterostructure; $NiTiO_3$; BiOBr

1. Introduction

Currently, water pollution has become an increasingly serious environmental problem. Organic pollutants in wastewater discharged from factories or hospitals are usually hazardous to humans and difficult to remove [1,2]. Solar-driven photocatalytic technology has been deemed an environmentally friendly and effective technology employed for solving problems [3,4]. Many photocatalysts have been confirmed to be able to degrade organic pollutants, including colored dyes and uncolored organics with stable structures [5], but their photocatalytic efficiency has not been satisfactory for practical needs yet.

The perovskite oxide, $NiTiO_3$, has recently attracted large interest as a photocatalyst for degrading toxic pollutants due to its visible-light harvesting ability (E_g = ~2.3 eV), good electronic transportability and high stability [6–9]. Nevertheless, the photocatalytic activity of single $NiTiO_3$ is inhibited by severe recombination of photogenerated carriers and narrow light absorption range [10]. One of the most effective methods to overcome the problems is combining $NiTiO_3$ with other semiconductors with matched band structures to form various heterojunctions [11,12]. Among them, the staggered type-II heterojunction could not only notably accelerate the separation of the carriers but also extend/enhance the light absorption range. For instance, Shi et al. [11] fabricated Type II junctions between 2D Cu_2WS_4 nanosheets and 1D $NiTiO_3$ nanofibers, which exhibited excellent degrading ability for RhB and TC because of improved visible-light adsorption, rapid separation of photogenerated carriers and large specific surface area. Wang et al. fabricated type-II junctions composed of 3D $NiTiO_3$ nanorods and 2D MoS_2 nanosheets, which also notably inhibited the recombination of electrons and holes and presented high photocatalytic H_2

production [12]. Consequently, constructing efficient type-II heterojunctions is an ideal path to improving the photocatalytic performance of $NiTiO_3$.

BiOBr is an efficient photocatalyst for degrading organic pollutants because of its strong oxidizing ability, lamellar structure and high chemical stability [13,14]. However, it suffers from the rapid combination of photoinduced carriers and poor visible-light absorption ($E_g \approx 2.7$ eV) [15]. Therefore, BiOBr was often employed to construct heterojunctions with the visible-light-responsive semiconductor to enhance the solar photocatalytic property of the composites such as BiOBr-Bi_2WO_6 [16], BiOBr/Bi_2MoO_6 [17] and WS_2/BiOBr [18], profiting from the efficient separation of photo-indued carriers and improved visible-light absorption efficiency. Supposed that BiOBr and $NiTiO_3$ have matched band structures, $NiTiO_3$-BiOBr heterojunctions would be expected to overcome their own shortcomings and exhibit high photocatalytic activity.

In addition, the photocatalytic behavior of the heterojunction also depends on the reasonable construction of the heterostructure, where there should be a large interface area and specific surface area for providing abundant active sites [19]. Consequently, employing the porous $NiTiO_3$ rods as supporters to in-situ grow the BiOBr nanosheets, the rod-like porous $NiTiO_3$-BiOBr binary heterojunctions were fabricated through precipitation, calcinations and hydrothermal methods for the first time in this study. Rhodamine B (RhB) and tetracycline (TC) were separately employed as objects of photodegradation to evaluate the performance of the composites. As a result, the composites presented highly improved photocatalytic performances compared to the single components. The photocatalyst undoubtedly exhibited high stability. Furthermore, a feasible photocatalytic mechanism was deduced based on free radical capture, EPR tests and band position measurements. This study would present a new way to construct stable $NiTiO_3$-based photocatalysts for degrading certain organics.

2. Experiments

2.1. Synthesis of $NiTiO_3$

$NiTiO_3$ was synthesized according to the method reported by Qu et al. [20], where 2.48 g of nickel acetate ($Ni(CH_3COO)_2 \cdot 4H_2O$) was dissolved into 60 mL of ethylene glycol (EG) under stirring, and then 3.4 mL of tetra-n-butyl titanate ($Ti(OC_4H_9)_4$) was added dropwise in sequence. During this process, light blue precipitates were produced, which gradually increased in quantity. The reaction continued for 1 h. Subsequently, the precipitates were collected, washed with deionized water and ethanol in turn, and dried at 80 °C. The resulting product was roasted in the air at 600 °C for 2 h to attain yellow $NiTiO_3$ samples.

2.2. Synthesis of $NiTiO_3$-BiOBr Composites

The $NiTiO_3$-BiOBr composites were prepared by hydrothermal treatment. First, 0.1 g $NiTiO_3$ was added to 30 mL of deionized water under stirring to achieve a homogeneous suspension. Second, 10 mL of EG containing an appropriate proportion of $Bi(NO_3)_3 \cdot 5H_2O$ was added dropwise with stirring and then continuously stirred for an extra 30 min. Subsequently, a certain volume of 1 $mg \cdot mL^{-1}$ KBr solution was also added to it and fully mixed. The resulted mixture was placed into a Teflon-lined stainless autoclave (100 mL) and kept heating at 180 °C for 12 h. As a result, the product was obtained by centrifugation, washed and dried as described above. The obtained products were named $NiTiO_3$-BiOBr(x), in which x signifies the designed weight percentage of BiOBr in the composite (5 wt%, 10 wt%, 15 wt% and 20 wt%, respectively). As a control experiment, a single BiOBr was synthesized with the same procedure as above without the addition of a $NiTiO_3$ sample.

2.3. Characterization

X-ray diffraction (XRD) spectra were employed to detect the crystalline properties of the products and performed on the XRD instrument (D/MAX-RB, Rigaku, Japan). A scanning electron microscope (SEM) equipped with an energy-dispersive X-ray (EDX) spectrometer (SU8010, Hitachi, Japan) was applied to determine the morphologies and elemental distribution on the surface of samples. A transmission electron microscope (TEM, F-20, FEI, USA) was employed to analyze the internal morphology of samples by observing TEM and high-resolution TEM (HRTEM) images. Pore structure was characterized using an instrument from Quantchrome NOVA 4200e (USA). To study the chemical components and oxidation states of elements in the materials, X-ray photoelectron spectra (XPS) were tested on an XPS instrument (EscaLab 250Xi, Thermo, USA). UV-visible diffuse reflectance spectra (UV-vis DRS) were applied to explore the light absorption of samples and carried out on a UV-vis spectrophotometer (T9s, Persee, China). To detect the separation efficiency of photogenerated carriers, the photoluminescence (PL) spectra were tested on a fluorescence spectrophotometer (F-7000, Hitachi, Japan). To determine the main free radicals in the photocatalytic system, an electron paramagnetic resonance (EPR) spectrometer (JES-FA200, JEOL, Japan) was employed to obtain the EPR spectra. An electrochemical workstation (5060F, RST, China) with a three-electrode cell system was applied to attain the Mott–Schottky plots to determine the flat band potentials of pristine semiconductors. A saturated calomel electrode (SCE) and Pt filament were employed as the reference and auxiliary electrodes, respectively. A certain amount of the sample was ultrasonically dispersed in the mixed solution of ethanol and Nafion and then coated into a circular trough on the ITO conductive glass, which was dried and used as the working electrode.

2.4. Photocatalytic Activity Experiments

The photocatalytic properties of the samples were evaluated by degrading RhB (20 mg·L^{-1}) and TC (40 mg·L^{-1}). A 400 W xenon lamp with a filter (>420 nm) was used as a light source. Typically, a 30 mg sample was added to 30 mL of an aqueous solution containing the organic pollutant under stirring. The suspension was kept in the dark for 60 min to reach the adsorption-desorption balance. After the lamp was turned on, 3.0 mL of the suspension was taken out and centrifuged at regular intervals (30 min for RhB and TC). The concentrations © of RhB and TC were measured at 554 nm and 357 nm, respectively, on the T9s spectrophotometer. The value of C_t/C_0 could signify degradation efficiency. Here, C_0 and C_t mean the concentration at the initial time and a certain time, respectively.

3. Results and Discussion

3.1. XRD Analysis

The XRD spectra of the samples are shown in Figure 1. The pattern of single $NiTiO_3$ is well in line with the standard pattern (JCPDS 33-0960). The characteristic peaks at 24.13°, 33.09°, 35.66°, 40.85°, 49.45°, 54.02°, 62.45° and 64.07° correspond to crystalline planes (012), (104), (110), (113), (024), (116), (214) and (300), respectively. In comparison, the patterns of $NiTiO_3$-BiOBr composites clearly show all these peaks. Besides, the peaks at 10.90°, 25.16°, 31.70°, 32.22°, 39.38°, 46.21° and 57.12° indexed to planes (001), (101), (102), (110), (112), (200) and (212) of BiOBr (JCPDS 09-0393), respectively, could be easily observed on the patterns of the composites with high BiOBr content (15% and 20%). With the decrease in BiOBr content, the intensity of these peaks gradually decreases or becomes invisible. These results suggest the successful synthesis of the $NiTiO_3$-BiOBr composite.

3.2. SEM and TEM Analysis

Taking the $NiTiO_3$-BiOBr (15%) composites as an example, the morphology of the prepared samples could be determined by SEM, TEM and HRTEM images. $NiTiO_3$ shows a rod-like shape with aspect ratio of ~3–5 (Figure 2a). It is also noted that a small number of irregular nanoparticles accumulates around the rods, which could be prompted by the growth of crystals during the roasting process [20]. Independent BiOBr is circular flakes with 150–300 nm in diameter (Figure 2b). The morphology of $NiTiO_3$ does not significantly change after combining with BiOBr (Figure 2c). From the TEM image (Figure 2d), it can be seen that the rod-like composite is made of many tiny nanoparticles and presents a porous structure, which should have a large specific surface area and interface area. In the HRTEM image (Figure 2e), the lattice spacing of 0.27 nm is attributed to the (104) lattice planes of $NiTiO_3$, and the lattice spacings of 0.23 and 0.18 nm correspond to the (112) and (104) planes of BiOBr sheets, respectively. This verifies the formation of the $NiTiO_3$-BiOBr heterojunction.

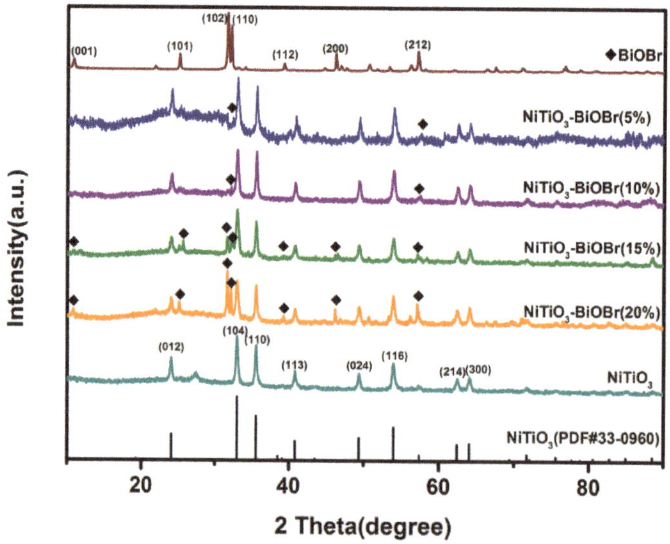

Figure 1. XRD patterns of the $NiTiO_3$, BiOBr and $NiTiO_3$-BiOBr composites.

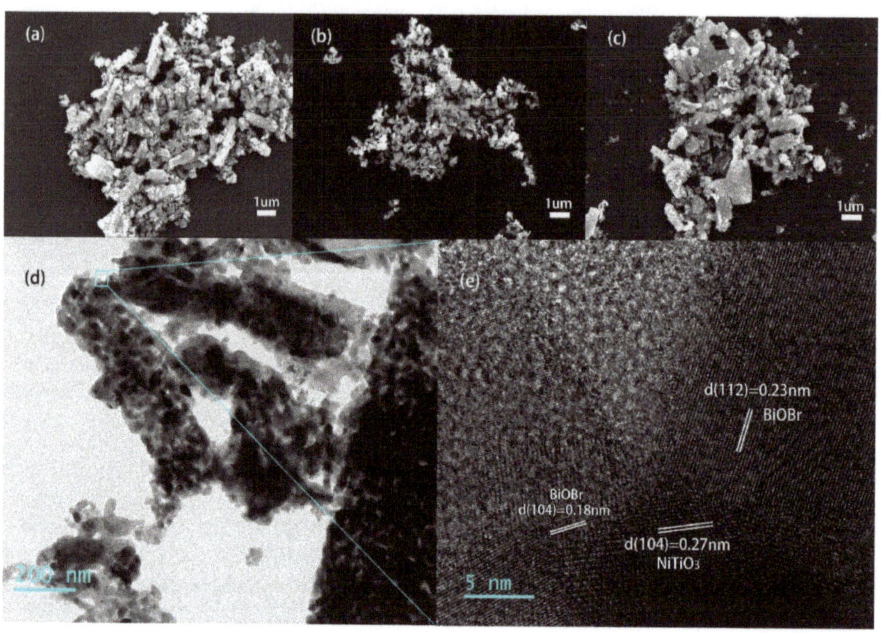

Figure 2. SEM pictures of (**a**) $NiTiO_3$, (**b**) BiOBr, and (**c**) $NiTiO_3$-BiOBr (15%) composites; (**d**) TEM image of $NiTiO_3$-BiOBr (15%); (**e**) HRTEM image of $NiTiO_3$-BiOBr (15%).

To illustrate the pore structure, N_2 adsorption isotherms and pore distributions of $NiTiO_3$ and $NiTiO_3$-BiOBr (15%) are depicted in Figure 3a,b. Obviously, they belong to meso- and macro-pore structures with pore sizes of 50–150 nm, which leads to their large specific surface areas (43.2 $m^2 \cdot g^{-1}$ for $NiTiO_3$, 33.0 $m^2 \cdot g^{-1}$ for the composite). The smaller surface area of the composite should result from the deposition of BiOBr, which blocked the partial pore of the $NiTiO_3$ rod. These results indicate that the composite has many abundant active sites and large interface area, which would be very beneficial for the photocatalytic process.

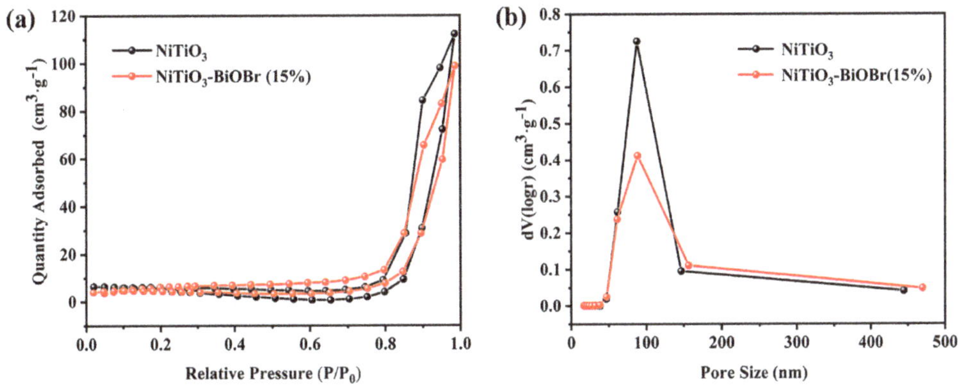

Figure 3. The N_2 adsorption isotherms (**a**) and pore distributions (**b**) of $NiTiO_3$ and $NiTiO_3$-BiOBr (15%).

3.3. XPS Analysis

The chemical composition and elemental oxidation states of the composites were further examined by XPS. The characteristic peaks of Ni, Ti, Bi, O and Br elements could be obviously detected on the full survey spectra of $NiTiO_3$-BiOBr (15%) (Figure 4a), indicating that these elements coexist in the binary composites. The spectrum of Ni 2p (Figure 4b) was fitted into two groups of peaks separately for both $NiTiO_3$ and binary composites, where the strong peaks at 855.33 and 872.72 eV are related to Ni $2p_{3/2}$ and Ni $2p_{1/2}$ of Ni^{2+}, respectively, and the other two weak peaks at 861.32 and 879.52 eV should be attributed to their satellite peaks [21]. Two obvious peaks of Ti 2p at 457.88 and 463.70 eV are assigned to the Ti $2p_{3/2}$ and Ti $2p_{1/2}$ of Ti^{4+} in $NiTiO_3$, respectively (Figure 4c) [21]. In Figure 4d,e, the peaks at 158.89 and 164.23 eV correspond to the Bi $4f_{7/2}$ and Bi $4f_{5/2}$ of Bi^{3+}, respectively, and the peaks at 67.58 and 68.78 eV ascribe to Br $3d_{5/2}$ and Br $3d_{3/2}$ of Br^-, separately [22,23], indicating the existence of the BiOBr component. For the O 1s spectra in Figure 4f, the peaks at 529.78 and 531.33 eV attribute to the O^{2-} of Bi-O bonds and adsorbed oxygen on the spectrum of single BiOBr, respectively [16], while the peaks at 529.65 and 531.20 eV designate the lattice O^{2-} in $NiTiO_3$ and the surface-adsorbed oxygen on the spectrum of $NiTiO_3$, respectively [24]. In comparison, the peak at 529.63 eV on the spectrum of the binary composites should be lattice O^{2-} in $NiTiO_3$ and BiOBr, while the peak at 530.86 eV could belong to the adsorbed oxygen. Apparently, the results further suggest the successful preparation of the $NiTiO_3$-BiOBr heterojunction.

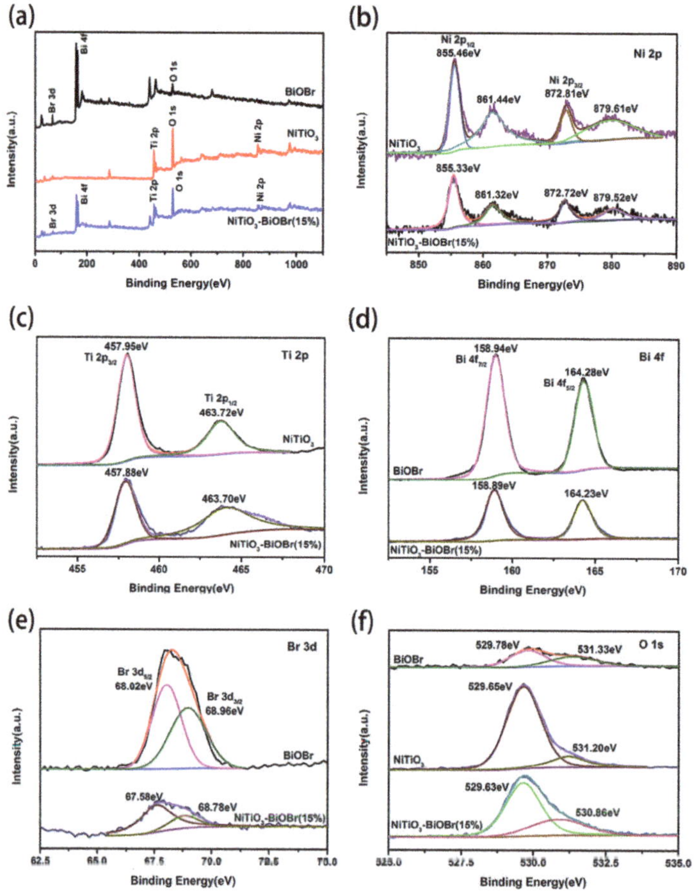

Figure 4. XPS analysis portraying $NiTiO_3$, BiOBr and $NiTiO_3$-BiOBr (15%): (**a**) Survey; (**b**) Ni 2p; (**c**) Ti 2p; (**d**) Bi 4f; (**e**) Br 3d; (**f**) O 1s.

3.4. Photocatalytic Properties

As shown in Figure 5a, only 14.6% and 49.0% of RhB within 1.5 h were degraded under visible-light irradiation over pristine $NiTiO_3$ and BiOBr, respectively. All binary materials, comparatively, exhibit higher degradation abilities for RhB. Among them, $NiTiO_3$-BiOBr (15%) presents the highest efficiency, i.e., 96.6% within 1.5 h (100% within 120 min), which is 6.61 times that of pure $NiTiO_3$. This indicates that the photocatalytic activity of $NiTiO_3$ in RhB degradation is amazingly boosted after coupling with an appropriate quantity of BiOBr, and the constructed $NiTiO_3$-BiOBr heterojunction is highly efficient. Compared with other reported $NiTiO_3$-based photocatalysts for degrading RhB, the $NiTiO_3$-BiOBr photocatalyst presents notably high degradation efficiency in 90 min even if high concentration of RhB was used in this work (Table 1). The photodegradation of TC over the materials was further tested. As shown in Figure 5b, the composites also exhibit much higher activity in degrading TC than pristine $NiTiO_3$ and BiOBr. The $NiTiO_3$-BiOBr (15%) composite also has the highest photodegradation rate (73.5%) within 3 h, which is far higher than $NiTiO_3$.

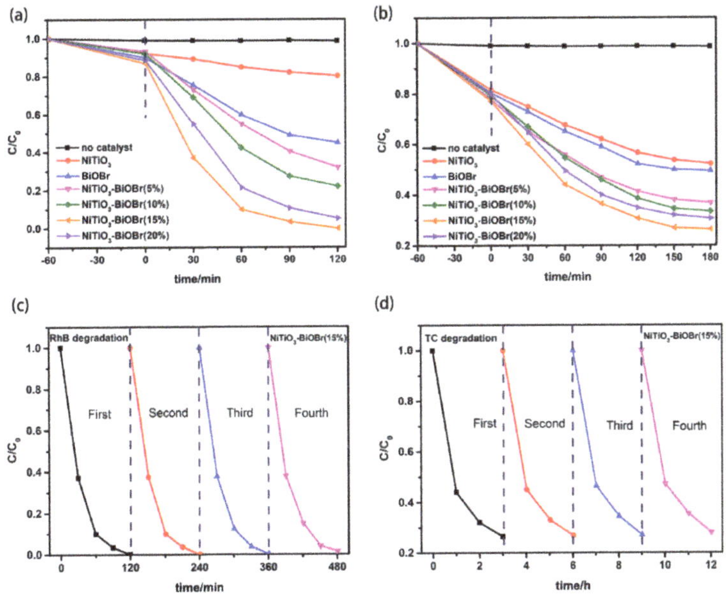

Figure 5. Photocatalytic degradation of prepared photocatalysts in degrading RhB (**a**) and TC (**b**); cycling tests of degrading RhB (**c**) and TC (**d**).

Cycling experiments over $NiTiO_3$-BiOBr (15%) for RhB and TC degradation were conducted to evaluate the stability of the materials. From Figure 5c,d, it is observed that the photocatalytic rate is basically unchanged after four repeated experiments, no matter what is degraded. Therefore, the stability of the binary composites is sufficient for the practical needs of photocatalytic systems.

Table 1. Comparison in the photodegradation performance for RhB among $NiTiO_3$-based photocatalysts.

Photocatalysts	C_0/mg·L^{-1}	Time/min	Dosage of Catalyst/g·L^{-1}	Efficiency/%	Refs.
$NiTiO_3$-Bi_4NbO_8Cl	5	90	1.0	~50	[24]
$NiTiO_3$-GO	10	90	0.2	~90	[25]
$NiTiO_3$-TiO_2	-	90	1.0	~62	[26]
$NiTiO_3$-Bi_2MoO_6	10	90	0.6	~92	[27]
$NiTiO_3$-α-Fe_2O_3	3	90	0.1	~75	[28]
$NiTiO_3$-ZnO	5 ppm	120	0.5	95	[29]
$NiTiO_3$-BiOBr	20	90	1.0	96.6	This work

3.5. Photocatalytic Mechanism Discussion

3.5.1. UV-Vis DRS Analysis

The UV-Vis DRS of the samples is employed to analyze their optical absorption. As shown in Figure 6a, $NiTiO_3$ exhibits absorption ability in both the ultraviolet and visible light ranges. An obvious absorption peak is observed near 450 nm. According to the literature [11,20,21], crystal field splitting of the Ni-O octahedron in $NiTiO_3$ crystal causes the $3d^8$ orbits of Ni^{2+} to split up into two sub-bands, and the Ni^{2+}/Ti^{4+} charge-transfer could lead to the formation of two adsorption peaks at ~450 and ~510 nm (not clear), respectively. In addition, $NiTiO_3$ also has light response above 600 nm due to the spin-allowing d-d transitions of Ni^{2+} [30]. In contrast, BiOBr mainly absorbs ultraviolet light and has an absorption edge of ~430 nm. By coupling them together, the composites present a notably strong visible-light absorption characteristic compared to pristine $NiTiO_3$ and BiOBr.

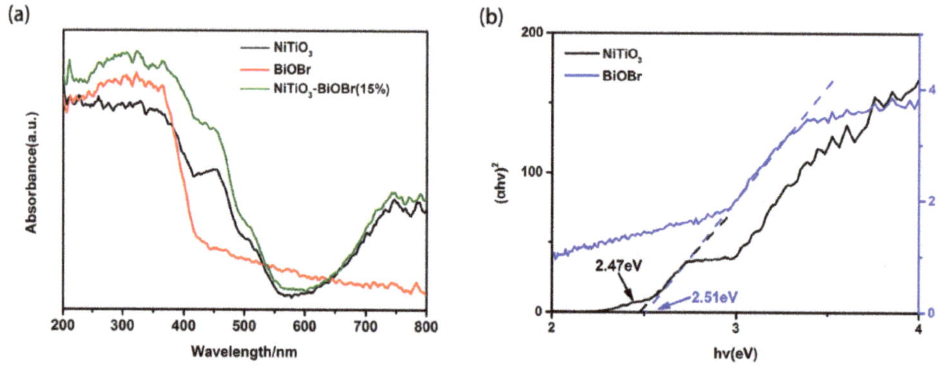

Figure 6. (**a**) UV-Vis DRS of $NiTiO_3$, BiOBr and $NiTiO_3$-BiOBr (15%) samples and (**b**) Tauc's plots of $NiTiO_3$ and BiOBr.

The band gaps of the two pristine samples were determined based on the equation: $(\alpha h\nu)^{2/n} = h\nu - E_g$, where α, h, ν, hν and E_g represent absorbance coefficient, plank constant, light frequency, irradiation energy and band gap energy, respectively [31]. Here, n is 1 for $NiTiO_3$ (direct bandgap) [26] and 4 for BiOBr (indirect bandgap) [32]. Therefore, the E_g of $NiTiO_3$ and BiOBr was separately determined to be 2.47 and 2.51 eV, respectively (Figure 6b).

3.5.2. PL Emission Spectra

The function of the heterojunctions in inhibiting the recombination of photogenerated carriers can be revealed by PL spectra. In Figure 7, the composites present much lower peak intensities than single $NiTiO_3$ and BiOBr, indicating that the formation of heterojunction greatly inhibited the recombination of photogenerated carriers and prompted the separation of the carriers. Among them, $NiTiO_3$-BiOBr (15%) presents the lowest peak intensity, implying its highest separation efficiency of carriers, which corresponds to its best photocatalytic performance.

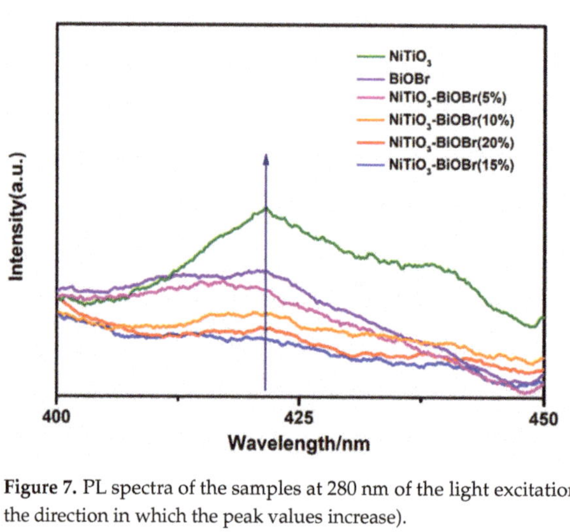

Figure 7. PL spectra of the samples at 280 nm of the light excitation wavelength (the arrow means the direction in which the peak values increase).

3.5.3. Free Radical Capture Tests

Hydroxyl radicals (·OH), superoxide radicals ($\cdot O_2^-$) and holes (h^+) are usually responsible for the degradation of organics. Free radical capture tests were performed for the $NiTiO_3$-BiOBr (15%) composite to investigate the degradation mechanisms of RhB and TC. $Na_2C_2O_4$ (10 mM), isopropanol (IPA, 10 mM) and benzoquinone (BQ, 1 mM) act as scavengers for h^+, ·OH and $\cdot O_2^-$, respectively. It is seen from Figure 8a that the photodegradation efficiency of RhB over $NiTiO_3$-BiOBr (15%) mildly decreases after adding IPA but significantly decreases after the addition of $Na_2C_2O_4$ or BQ, implying that h^+ and $\cdot O_2^-$ are the main free radicals, while ·OH has little contribution to the RhB degradation. For TC (Figure 8b), the degradation drastically decreases to near zero after adding BQ and also clearly declines when IPA is added, while there is little change after adding $Na_2C_2O_4$. This demonstrates that $\cdot O_2^-$ and ·OH play key roles in the following order: $\cdot O_2^- > \cdot OH$, while h^+ hardly contributed to the degradation of TC, possibly due to its relatively low oxidability.

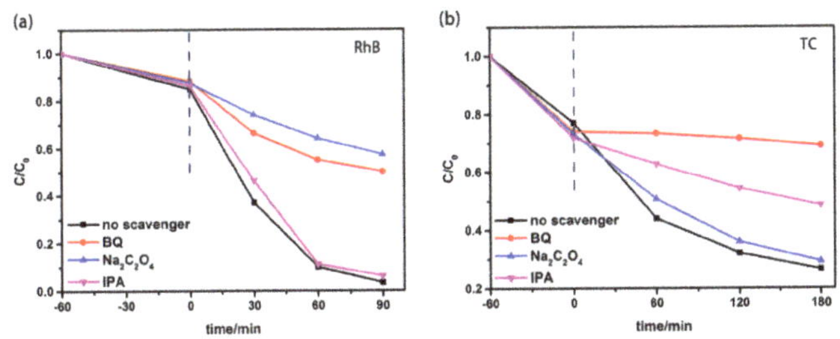

Figure 8. Degradation of RhB (**a**) and TC (**b**) over $NiTiO_3$-BiOBr (15%) with different scavengers under visible-light irradiation.

EPR measurements were performed to further explore the change of free radicals with the formation of heterojunctions. In Figure 9a, the ·OH signal in the $NiTiO_3$-BiOBr composite system is hardly observed, while a weak signal could be detected in the single $NiTiO_3$ system, indicating that the heterojunction does not promote or is adverse to the formation of ·OH. In contrast, the DMPO-$\cdot O_2^-$ signals can be explicitly detected. Additionally, the composite system exhibits a stronger signal than the single $NiTiO_3$ system (Figure 9b). This suggests that the heterojunction facilitates $\cdot O_2^-$ production, which is consistent with the results of the capture tests and show that $\cdot O_2^-$ plays a key role in the photocatalytic process.

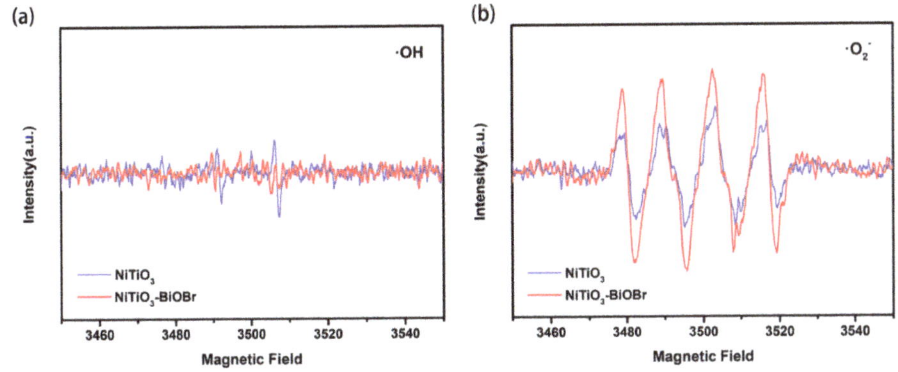

Figure 9. EPR spectra of (**a**) DMPO-·OH and (**b**)$\cdot O_2^-$ over $NiTO_3$ and $NiTiO_3$-BiOBr composite with visible-light irradiation.

3.5.4. Band Position Determination

The band positions of $NiTiO_3$ and BiOBr were determined by both Mott–Schottky (MS) plots and their band gap energies. As shown in Figure 10, both $NiTiO_3$ and BiOBr belong to n-type semiconductors, as their MS plots exhibit positive slopes. The flat band energy (E_{fb}) of $NiTiO_3$ and BiOBr could be determined as −0.48 eV and −0.89 eV (vs. SCE), respectively, based on their intersections of the X-axis. The conduction band (CB) energy is usually ~0.1 eV more negative than E_{fb} for n-type semiconductors [33]. Consequently, the E_{CB} of $NiTiO_3$ and BiOBr are identified as −0.34 eV and −0.75 eV vs. NHE, respectively, according to the equation: NHE = 0.24 + SCE. Based on the E_g of $NiTiO_3$ (2.47 eV) and BiOBr (2.51 eV), the E_{VB} of $NiTiO_3$ and BiOBr are 2.13 eV and 1.76 eV ($E_{VB} = E_{CB} - E_g$), respectively.

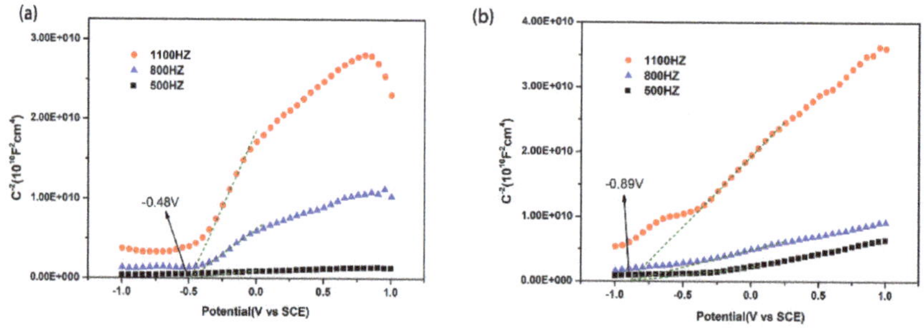

Figure 10. Mott–Schottky plots of (**a**) $NiTiO_3$ and (**b**) BiOBr.

3.6. *Photocatalytic Mechanism of $NiTiO_3$-BiOBr Photocatalysts*

Based on the above analysis, the transfer pathway of photoexcited carriers on the interface of the $NiTiO_3$-BiOBr heterojunction was discussed to clarify the photocatalytic degradation mechanism. As shown in Figure 11, electrons (e^-) on the VB of both $NiTiO_3$ and BiOBr would separately attain the corresponding light energy to jump to their CB under visible-light irradiation, leaving holes (h^+) on their VB. According to the CB and VB positions of the two semiconductors, the photo-induced carrier migration could follow the S-scheme transfer pathway (Figure 11a) [34,35] or type II transfer pathway (Figure 11b) [11,12]. Here, if the carriers follow the former, the e^- on the CB of $NiTiO_3$ would combine with the h^+ on the VB of BiOBr, thus inhibiting the recombination of e^- on CB of BiOBr and h^+ on the VB of $NiTiO_3$ and remaining their inherent redox capacity. This means that accumulating h^+ on the VB of $NiTiO_3$ in the composites could oxidize OH^- to generate more $\cdot OH$ than that in the single $NiTiO_3$. However, it is conflicted with the experimental results of the EPR (Figure 9a). Consequently, the S-scheme transfer pathway is excluded, and the conventional type II transfer pathway is proposed.

As illustrated in Figure 11b, the potential level difference between BiOBr and $NiTiO_3$ would lead to the transfer of e^- from BiOBr to $NiTiO_3$ and the transfer of h^+ in the opposite direction, resulting in the highly efficient separation of the carriers. The e^- accumulated on the CB of $NiTiO_3$ would reduce the adsorbed oxygen to generate a large quantity of $\cdot O_2^-$ due to their more negative potential than $E^0(O_2/\cdot O_2^-)$ (0.13 eV vs. NHE) [36], efficiently degrading RhB and TC. The accumulating h^+ on the VB of BiOBr could directly oxidize RhB. Nevertheless, they could not oxidize OH^- to $\cdot OH$ according to their lower potential than E^0 ($\cdot OH/OH^-$) (1.99 eV vs. NHE) [24], which is well in line with the EPR results. Consequently, the composites exhibited notably improved photodegradation efficiency for RhB and TC due to the high separation of photogenerated carriers benefiting from the type II heterojunctions with large interface areas. Additionally, the strengthened visible light absorption of the composites also causes an improvement in their photocatalytic efficiency.

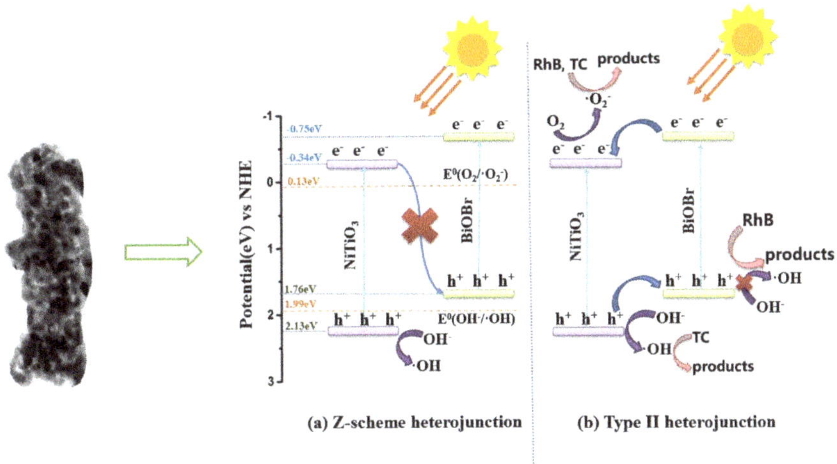

Figure 11. Proposed charge transfer pathway on the interface of heterojunction and photocatalytic mechanism of the $NiTiO_3$-BiOBr composites in degrading organic pollutants.

4. Conclusions

In this study, a series of novel binary $NiTiO_3$-BiOBr materials were synthesized via precipitation, calcination and hydrothermal methods in turn. As shown by various characterizations, BiOBr nanosheets were decorated on the $NiTiO_3$ nanoparticals, forming porous rod-like $NiTiO_3$-BiOBr heterojunctions with large interface area. The visible-light photodegradation capacity of the binary composites was significantly boosted compared with single $NiTiO_3$ and BiOBr. The optimum composite degraded 96.6% of RhB within 1.5 h (100% in 2 h) and 73.5% of TC within 3 h, which are 6.61 and 1.53 times those of $NiTiO_3$, respectively. In addition, it exhibited excellent photostability. The UV-Vis DRS spectroscopy and PL analysis indicate that the $NiTiO_3$-BiOBr heterojunctions strengthened visible-light absorption and notably facilitated the separation of photogenerated carriers, resulting in high photocatalytic activity. Considering the results of the free radical capture tests, EPR measurements and band position analysis, a feasible photodegradation mechanism of type-II heterojunction was proposed. This work provided a strategy for fabricating highly efficient $NiTiO_3$-based photocatalysts for degrading certain organic objects.

Author Contributions: Conceptualization, K.S.; Methodology, K.S.; Formal analysis, H.Z. and X.M.; Investigation, K.S. and M.L.; Resources, X.M. and W.L.; Data curation, M.L.; Writing—original draft, K.S.; Writing—review & editing, K.S. and H.Z.; Project administration, W.L. All authors have read and agreed to the published version of the manuscript.

Funding: The financial support provided by the Open Research Fund of State Key Laboratory of Multiphase Complex Systems (MPCS-2021-D-11).

Institutional Review Board Statement: Not applicable.

Informed Consent Statement: Not applicable.

Data Availability Statement: The authors confirm that the data supporting the findings of this study are available within the article.

Acknowledgments: We gratefully acknowledge the financial support provided by the Open Research Fund of State Key Laboratory of Multiphase Complex Systems (MPCS-2021-D-11).

Conflicts of Interest: The authors declare no conflict of interest.

References

1. Sun, K.Y.; Zhou, H.L.; Li, X.Y.; Ma, X.H.; Zhang, D.H.; Li, M.C. The novel 2-dimensional Bi_2MoO_6-Bi_2O_3-Ag_3PO_4 ternary photocatalystwith n-n-p heterojunction for enhanced degradation performance. *J. Alloys Compd.* **2022**, *913*, 165119. [CrossRef]
2. Ibrahim, I.; Belessiotis, G.; Antoniadou, M.; Kaltzoglou, A.; Sakellis, E.; Katsaros, F.; Sygellou, L.; Arfanis, M.; Salama, T.; Falaras, P. Silver decorated TiO_2/g-C_3N_4 bifunctional nanocomposites for photocatalytic elimination of water pollutants under UV and artificial solar light. *Results Eng.* **2022**, *14*, 100470. [CrossRef]
3. Fu, K.; Pan, Y.; Ding, C.; Shi, J.; Deng, H. Photocatalytic degradation of naproxen by Bi_2MoO_6/g-C_3N_4 heterojunction photocatalyst under visible light: Mechanisms, degradation pathway, and DFT calculation. *J. Photochem. Photobiol. A Chem.* **2021**, *412*, 113235. [CrossRef]
4. Li, Q.Q.; Zhao, W.L.; Zhai, Z.C.; Ren, K.X.; Wang, T.Y.; Guan, H.; Shi, H.F. 2D/2D Bi_2MoO_6/g-C_3N_4 S-scheme heterojunction photocatalyst with enhanced visible-light activity by Au loading. *J. Mater. Sci. Technol.* **2020**, *56*, 216–226. [CrossRef]
5. Falara, P.; Ibrahim, I.; Zourou, A.; Sygellou, L.; Sanchez, D.E.; Romanos, G.; Givalou, L.; Antoniadou, M.; Arfanis, M.; Han, C.; et al. Bi-functional photocatalytic heterostructures combining titania thin films with carbon quantum dots (C-QDs/TiO_2) for effective elimination of water pollutants. *Environ. Sci. Pollut. Res. Int.* **2023**, 1–16. [CrossRef] [PubMed]
6. Li, Y.B.; Wang, G.R.; Wang, Y.B.; Jin, Z.L. Phosphating 2D CoAl LDH anchored on 3D self-assembled $NiTiO_3$ hollow rods for efficient hydrogen evolution. *Catal. Sci. Technol.* **2020**, *10*, 2931–2947. [CrossRef]
7. Liu, J.; Li, X.H.; Han, C.H.; Zhou, X.J.; Li, X.W.; Liang, Y.; Liu, S.; Shao, C.L.; Liu, Y.C. Ternary $NiTiO_3$@g-C_3N_4-Au nanofibers with a synergistic Z-scheme core@shell interface and dispersive Schottky contact surface for enhanced solar photocatalytic activity. *Mater. Chem. Front.* **2021**, *5*, 2730–2741. [CrossRef]
8. Trang, N.T.T.; Khang, D.M.; Dung, D.D.; Trung, N.N.; Phuong, N.T.; Bac, L.H. Synthesis of Ilmenite $NiTiO_3$ Rods and Effect of pH on Rhodamine B Textile Dye Degradation under LED Visible-Light Irradiation. *J. Electeon. Mater.* **2021**, *50*, 7188–7197. [CrossRef]
9. Absalan, Y.; Bratchikova, I.; Kovalchukova, O. Accurate investigation to determine the best conditions for using $NiTiO_3$ for bromophenol blue degradation in the environment under UV–vis light based on concentration reduction and to compare it with TiO_2. *Environ. Nanotechnol. Monit. Manag.* **2017**, *8*, 244–253. [CrossRef]
10. Wang, H.; Yuan, X.Z.; Wang, H.; Chen, X.H.; Wu, Z.B.; Jiang, L.B.; Xiong, W.P.; Zhang, Y.X.; Zeng, G.M. One-step calcination method for synthesis of mesoporous g-C_3N_4/$NiTiO_3$ heterostructure photocatalyst with improved visible light photoactivity. *RSC Adv.* **2015**, *5*, 95643–95648. [CrossRef]
11. Peng, D.X.; Wang, Y.T.; Shi, H.F.; Jiang, W.; Jin, T.; Jin, Z.H.; Chen, Z. Fabrication of novel Cu_2WS_4/$NiTiO_3$ heterostructures for efficient visible-light photocatalytic hydrogen evolution and pollutant degradation. *J. Colloid Interface Sci.* **2022**, *613*, 194–206. [CrossRef] [PubMed]
12. Li, H.Y.; Wang, G.R.; Gong, H.M.; Jin, Z.L. Phosphated 2D MoS_2 nanosheets and 3D $NiTiO_3$ nanorods for efficient photocatalytic hydrogen evolution. *ChemCatChem* **2020**, *12*, 5492–5503. [CrossRef]
13. Li, Z.Z.; Zhang, H.G.; Wang, L.; Meng, X.C.; Shi, J.J.; Qi, C.X.; Zhang, Z.S.; Feng, L.J.; Li, C.H. 2D/2D BiOBr/Ti_3C_2 heterojunction with dual applications in both waterdetoxification and water splitting. *J. Photochem. Photobiol. A Chem.* **2020**, *386*, 112099. [CrossRef]
14. Li, X.B.; Xiong, J.; Gao, X.M.; Ma, J.; Chen, Z.; Kang, B.B.; Liu, J.Y.; Li, H.; Feng, Z.J.; Huang, J.T. Novel BP/BiOBr S-scheme nano-heterojunction for enhanced visible-light photocatalytic tetracycline removal and oxygen evolution activity. *J. Hazard. Mater.* **2020**, *387*, 121690. [CrossRef]
15. Liu, H.; Zhou, H.L.; Liu, X.T.; Li, H.D.; Ren, C.J.; Li, X.Y.; Li, W.J.; Lian, Z.Q.; Zhang, M. Engineering design of hierarchical g-C_3N_4@Bi/BiOBr ternary heterojunction with Z-scheme system for efficient visible-light photocatalytic performance. *J. Alloys Compd.* **2019**, *798*, 741–749. [CrossRef]
16. Liu, K.; Zhang, H.B.; Muhammad, Y.; Fu, T.; Tang, R.; Tong, Z.F.; Wang, Y. Fabrication of n-n isotype BiOBr-Bi_2WO_6 heterojunctions by inserting Bi_2WO_6 nanosheets onto BiOBr microsphere for the superior photocatalytic degradation of Ciprofloxacin and tetracycline. *Sep. Purif. Technol.* **2021**, *274*, 118992–119005. [CrossRef]
17. Hao, L.; Ju, P.; Zhang, Y.; Sun, C.J.; Dou, K.P.; Liao, D.K.; Zhai, X.F.; Lu, Z.X. Novel plate-on-plate hollow structured BiOBr/Bi_2MoO_6 p-n heterojunctions: In-situ chemical etching preparation and highly improved photocatalytic antibacterial activity. *Sep. Purif. Technol.* **2022**, *298*, 121666–121682. [CrossRef]
18. Fu, S.; Yuan, W.; Liu, X.M.; Yan, Y.H.; Liu, H.P.; Li, L.; Zhao, F.Y.; Zhou, J.G. A novel 0D/2D WS_2/BiOBr heterostructure with rich oxygen vacancies for enhanced broad-spectrum photocatalytic performance. *J. Colloid Interface Sci.* **2020**, *569*, 150–163. [CrossRef]
19. Yan, Q.S.; Guo, Z.Y.; Wang, P.Y.; Cheng, Y.N.; Wu, C.Y.; Zuo, H.R. Facile construction of 0D/2D In_2O_3/Bi_2WO_6 Z-scheme heterojunction with enhanced photocatalytic activity for antibiotics removal. *J. Alloys Compd.* **2023**, *937*, 168362. [CrossRef]
20. Qu, Y.; Zhou, W.; Ren, Z.Y.; Du, S.C.; Meng, X.Y.; Tian, G.H.; Pan, K.; Wang, G.F.; Fu, H.G. Facile preparation of porous $NiTiO_3$ nanorods with enhanced visible-light-driven photocatalytic performance. *J. Mater. Chem.* **2012**, *22*, 16471–16476. [CrossRef]
21. Kim, S.R.; Jo, W.K. Application of a photostable silver-assisted Z-scheme $NiTiO_3$ nanorod/g-C_3N_4 nanocomposite for efficient hydrogen generation. *Int. J. Hydrogen Energy* **2019**, *44*, 801–808. [CrossRef]
22. Shi, Y.Q.; Xiong, X.Y.; Ding, S.P.; Liu, X.F.; Jiang, Q.Q.; Hu, J.C. In-situ topotactic synthesis and photocatalytic activity of plate-like BiOCl/2D networks Bi_2S_3 heterostructures. *Appl. Catal. B Environ.* **2018**, *220*, 570–580. [CrossRef]
23. Liu, H.J.; Wang, B.J.; Chen, M.; Zhang, H.; Peng, J.B.; Ding, L.; Wang, W.F. Simple synthesis of BiOAc/BiOBr heterojunction composites for the efficient photocatalytic removal of organic pollutants. *Sep. Purif. Technol.* **2021**, *261*, 118286. [CrossRef]

24. Qu, X.F.; Liu, M.H.; Zhang, W.X.; Sun, Z.; Meng, W.; Shi, L.; Du, F.L. A facile route to construct $NiTiO_3/Bi_4NbO_8Cl$ heterostructures for enhanced photocatalytic water purification. *J. Mater. Sci.* **2020**, *55*, 9330–9342. [CrossRef]
25. Pham, T.T.; Nguyen-Huy, C.; Shin, E.W. $NiTiO_3$/reduced graphene oxide materials synthesized by a two-step microwave-assisted method. *Mater. Lett.* **2016**, *184*, 38–42. [CrossRef]
26. Lakhera, S.K.; Hafeez, H.Y.; Veluswamy, P.; Ganesh, V.; Khan, A.; Ikeda, H.; Neppolian, B. Enhanced photocatalytic degradation and hydrogen production activity of in situ grown TiO_2 coupled $NiTiO_3$ nanocomposites. *Appl. Surf. Sci.* **2018**, *449*, 790–798. [CrossRef]
27. Jin, Y.M.; Shen, X.F.; Liu, Z.X.; Wang, Z.J.; Zhu, B.; Xu, P.F.; Luo, L.; Zhang, L.S. Synthesis of $NiTiO_3$-Bi_2MoO_6 core-shell fiber-shaped heterojunctions as efficient and easily recyclable photocatalysts. *New J. Chem.* **2018**, *42*, 411–419. [CrossRef]
28. Zhang, Y.; Gu, J.; Murugananthan, M.; Zhang, Y.R. Development of novel α-$Fe_2O_3/NiTiO_3$ heterojunction nanofibers material with enhanced visible-light photocatalytic performance. *J. Alloys Compd.* **2015**, *630*, 110–116. [CrossRef]
29. Zarazúa-Morín, M.E.; Galindo-Luna, A.S.; Gallegos-Sánchez, V.J.; Zermeño-Resendiz, B.B.; Torres-Martínez, L.M. Novel hydrothermal-assisted microwave synthesis of $NiTiO_3$/ZnO and sonophotocatalytic effect for degradation of rhodamine B. *Top. Catal.* **2022**, *65*, 1182–1190. [CrossRef]
30. Wang, Z.Y.; Peng, J.W.; Feng, X.; Ding, Z.X.; Li, Z.H. Wide spectrum responsive CdS/$NiTiO_3$/CoS with superior photocatalytic performance for hydrogen evolution. *Catal. Sci. Technol.* **2017**, *7*, 2524–2530. [CrossRef]
31. Wang, Z.L.; Huo, Y.; Zhang, J.F.; Lu, C.; Dai, K.; Liang, C.H.; Zhu, G.P. Facile preparationof two-dimensional Bi_2MoO_6@Ag_2MoO_4 core-shell composite with enhancedvisible light photocatalytic activity. *J. Alloys Compd.* **2017**, *729*, 100–108. [CrossRef]
32. Hu, T.P.; Yang, Y.; Dai, K.; Zhang, J.F.; Liang, C.H. A novel Z-scheme Bi_2MoO_6/BiOBr photocatalyst for enhanced photocatalytic activity under visible light irradiation. *Appl. Surf. Sci.* **2018**, *456*, 473–481. [CrossRef]
33. Li, H.Y.; Wang, G.R.; Gong, H.M.; Jin, Z.L. Hollow Nanorods and Amorphous Co_9S_8 Quantum Dots Construct S-Scheme Heterojunction for Efficient Hydrogen Evolution. *J. Phys. Chem. C* **2021**, *125*, 648–659. [CrossRef]
34. Li, B.F.; Wang, W.J.; Zhao, J.W.; Wang, Z.Y.; Su, B.; Hou, Y.D.; Ding, Z.X.; Ong, W.J.; Wang, S.B. All-solid-state direct Z-scheme $NiTiO_3/Cd_{0.5}Zn_{0.5}S$ heterostructures for photocatalytic hydrogen evolution with visible light. *J. Mater. Chem. A* **2021**, *9*, 10270–10276. [CrossRef]
35. Pham, T.T.; Shin, E.W. Inhibition of charge recombination of $NiTiO_3$ photocatalyst by the combination of Mo-doped impurity state and Z-scheme charge transfer. *Appl. Surf. Sci.* **2020**, *501*, 143992. [CrossRef]
36. Tang, C.N.; Liu, E.Z.; Fan, J.; Hu, X.Y.; Ma, Y.N.; Wan, J. Graphitic-C_3N_4-hybridized Ag_3PO_4 tetrahedron with reactive{111} facets to enhance the visible-light photocatalytic activity. *RSC Adv.* **2015**, *5*, 91979–91987. [CrossRef]

materials

MDPI

Article

Enhanced Catalytic Activity and Energy Savings with Ni-Zn-Mo Ionic Activators for Hydrogen Evolution in Alkaline Electrolysis

Ivana Perović [1,*], Milica Marčeta Kaninski [2,*], Gvozden Tasić [1], Sladjana Maslovara [2], Petar Laušević [1], Mina Seović [1] and Vladimir Nikolić [2]

1 Department of Physical Chemistry, "Vinča" Institute of Nuclear Sciences—National Institute of the Republic of Serbia, University of Belgrade, Mike Petrovića Alasa 12-14, 11351 Belgrade, Serbia; gvozdentasic@vin.bg.ac.rs (G.T.); petarl@vin.bg.ac.rs (P.L.); mina050@vin.bg.ac.rs (M.S.)

2 Institute of General and Physical Chemistry, Studentski trg 12/V, 11158 Belgrade, Serbia; smaslovara@iofh.bg.ac.rs (S.M.); vnikolic@iofh.bg.ac.rs (V.N.)

* Correspondence: ivanaperovic@vin.bg.ac.rs (I.P.); milica@iofh.bg.ac.rs (M.M.K.)

Abstract: Green hydrogen produced by alkaline electrolysis is a promising solution to address the world's increasing energy demand while mitigating greenhouse gas emissions. However, the efficient and cost-effective production of green hydrogen via alkaline electrolysis requires improvements. This paper presents an in situ activation process that simplifies the alkaline electrolysis technology while enhancing the catalytic activity of electrodes for the hydrogen evolution reaction. The aim of this research is to enhance the energy efficiency of alkaline electrolysis and decrease the energy consumption for hydrogen production. To achieve this goal, ionic activators comprising Ni, Zn, and Mo were incorporated into the standard electrolyte solution. Our results demonstrate that the anticipated improvement in the catalytic activity of the d-metal combination, surpassing even that of precious metals, has been successfully attained. As a result, a 20% reduction in energy consumption (REC) for the hydrogen produced has been observed. The catalytic activity of the added activators for the hydrogen evolution reaction was discussed by analyzing the mechanism of the reaction via Tafel analysis and EIS techniques. These findings offer a promising approach to improve alkaline electrolysis and enhance the production of green hydrogen.

Keywords: hydrogen evolution reaction; alkaline electrolysis; energy saving; ionic activators; catalysis

1. Introduction

The efficient storage of energy obtained from intermittent renewable energy sources is crucial for improving these technologies. Storing surplus energy during periods of peak production guarantees a steady supply during times when renewable sources are inaccessible [1]. The most promising solution to this problem is the hydrogen economy, the backbone of which is hydrogen and hydrogen energy. The hydrogen economy is the most viable solution to this issue, as it is founded on hydrogen and its energy. The idea of a hydrogen-based economy was formulated as an alternative to fossil fuel-based economies as early as 1975 [2], but significant research toward improving technologies related to this concept has only been conducted in the last two decades.

The effective implementation of the hydrogen-based economy is heavily dependent on the cost of green hydrogen in the global market [3]. The production of hydrogen that is cost-efficient, energy-efficient, and has a low carbon footprint is crucial to meet market demands. To achieve this, reducing the energy losses of the hydrogen production process is key [4]. Recent progress in electrolysis technology has led to enhanced energy efficiency of the process by decreasing the initial production expenses and minimizing energy consumption and loss [5,6]. Any advancements in improving the efficiency of the electrolytic process hold great significance both scientifically and practically.

The electrolysis of water is a crucial method for hydrogen production, as it yields highly pure hydrogen that can be directly utilized in applications requiring high-purity hydrogen, such as fuel cells, without the need for additional purification steps. Moreover, when renewable energy sources like solar, wind, or hydro power are employed for the electrolysis process, it can significantly reduce or even eliminate environmental pollution.

There are various approaches to enhance the effectiveness of splitting water molecules into their constituents during the electrolysis process. These methods mainly involve improving the technical aspects of the electrolyzer, such as the geometry of the electrolytic cell, as well as developing new membranes, electrodes, and electrocatalytic materials [7]. This paper focuses on exploring the catalytic activity of different metals for the hydrogen evolution reaction (HER) in the context of the hydrogen evolution reaction. The catalytic activity of diverse metals is a periodic function of the atomic number, which is confined to the three periods of transition elements. The modulation in current density is indicative of the corresponding catalytic activity. The larger the exchange current density, the faster the reaction, and vice versa. Values of exchange current densities for HER as a function of the metal–hydrogen bonding energy ("volcano" curve) give us an insight into which metals are most suitable for the catalysis of the hydrogen evolution reaction. Considering that noble metals are at the top of the "volcano" curve, due to their expensive exploitation and application, researchers shifted their focus on finding new materials that will have similar or equal catalytic activity as noble ones, and at the same time, their application will be significantly more economical and available [8]. Brewer theory of intermetallic bonds tells us that the best ways to improve the efficiency of the electrolytic hydrogen evolution process is the presence of d-metals [9,10]. It predicts a synergistic effect of d-elements with hypo- and hyper-electronic configuration. The pairing of these elements creates a new phase that occupies some intermediate electronic configuration and is comparable to the configuration of individual transition metals. Taking this into account, as well as the fact that d orbitals participate in the process of adsorption of the reaction intermediate (for HER it is H_{ads}), it is expected that alloys of two or more elements from the opposite sides of the "volcano" curve show a higher activity than they would individually. In some cases, it is possible that the activity of such alloys exceeds the activity of precious metals. The real application of this synergistic effect in electrolysis is based on the modification of the surfaces of nickel or stainless-steel electrodes with transition metal alloys. The method by which these modifications are achieved is the most common electrochemical deposition of transition metals from appropriate electrodeposition baths [11–13]. Recently, a new method is in use, which involves the direct addition of metals (so-called ionic activators, i.a.) to the electrolyte during the electrolysis process itself [14]. This method is called in situ ionic activation. It implies the presence of the ionic form of the d-metal complex in the electrolyte solution, which reduces the overvoltage for hydrogen extraction and enables a more energetically favorable transfer of protons from the electrolyte to the cathode surface.

In our previous studies, we have substantiated the exceptional in situ catalytic efficacy exhibited by different combinations of d-metals for the hydrogen evolution reaction. Particularly noteworthy among these combinations is the Ni-Co-Mo system [15]. Building upon these findings, the purpose of the present investigation is to explore the catalytic performance of a Ni-Zn-Mo combination of d-metals. This choice stems from the fact that zinc (Zn) is a more accessible and cost-effective element compared to cobalt (Co), rendering it highly significant for potential industrial applications. The HER mechanism was analyzed under various experimental conditions, including different temperatures and applied current densities. Additionally, the study examined the relationship between the concentration of ionic activators, operating temperature, exchange current density, and energy consumption during in situ activation.

2. Materials and Methods

The experiments were conducted in a 6M (30 wt%) KOH solution (Merck KGaA, Darmstadt, Germany, p.a.) that was prepared using deionized water (with a resistivity of

18 $M\Omega cm^{-1}$). Salts and complex used as ionic activators: Sodium molybdate dihydrate ($Na_2MoO_4{\cdot}2H_2O$), Zinc chloride ($ZnCl_2$) and Tris(ethylenediamine)nickel(II) chloride dihydrate ($[Ni(en)_3]Cl_2{\cdot}2H_2O$) were all ACS reagents (Merck KGaA, Darmstadt, Germany). Table 1 shows the composition of the electrolytes that were used in the experiments that investigated the influence of ionic activators in detail. According to Brewers theory of synergistic effect [16] of d-metals and based on experience from our previous research [17,18], we tested different ratios of Ni and Zn with constant concentration of the Mo salt. These concentrations are optimized for our experimental setup. The active surfaces of all electrodes used in this work were made of high-purity nickel (99.99%). Preceding each series of measurements, the electrodes were subjected to a rigorous cleaning procedure, involving polishing with 800 and 2000 grit polishing paper, which was followed by immersion in a HNO_3 aqueous solution (molar ratio of 2:1) for a duration of 2 min and subsequent rinsing with deionized water and ethanol. The reference electrode was an SCE electrode, while a Pt mesh with a surface area considerably larger than that of the working electrodes under scrutiny was utilized as the counter electrode.

Table 1. Composition of investigated electrolytes with ionic activators.

Solution No.	c(KOH)/M	$c([Ni(en)_3]Cl_2{\cdot}2H_2O)$/M	$c(ZnCl_2)$/M	$c(Na_2MoO_4{\cdot}2H_2O)$/M
1	6	1×10^{-3}	1×10^{-3}	1×10^{-2}
2	6	1×10^{-3}	5×10^{-3}	1×10^{-2}
3	6	1×10^{-3}	1×10^{-2}	1×10^{-2}
4	6	5×10^{-3}	1×10^{-3}	1×10^{-2}
5	6	1×10^{-2}	1×10^{-3}	1×10^{-2}
6	6	5×10^{-2}	1×10^{-3}	1×10^{-2}

To investigate the impact of Ni-Zn-Mo activators on the kinetics of the hydrogen reaction, polarization curves were recorded using a three-compartment standard electrochemical cell. The current density range investigated was from 10 to 100 mA cm^{-2}. The electrode potential was altered in the range of −1.15 to −0.75 V, relative to the SCE reference electrode, at a rate of 1 mVs^{-1}. To correct for the potential drop across the electrolyte, the "current interrupt" method was employed. The recording of impedance spectra was performed in the frequency range from 0.1 to 100 kHz, using an alternating signal of amplitude 10 mV superimposed on a constant value of overvoltage in the range from −20 to −250 mV. Preceding every measurement, a constant potential was applied to the working electrode for a duration of 15 s to condition it. The temperature of the electrolyte solution within the electrochemical cell ranged between 298 and 343 K. Prior to each measurement, the electrolyte solution was thoroughly purged with hydrogen by continuous bubbling for a duration of 30 min. Due to the rapid kinetics associated with the hydrogen evolution reaction, it is imperative to conduct preliminary measurements following hydrogen bubbling. This precaution ensures that the evolution of hydrogen has a negligible impact on the local concentration, particularly in comparison to other gases such as nitrogen or argon. The Gamry Reference 3000 Potentiostat/Galvanostat/ZRA instrument (Gamry Instruments, Warminster, PA, USA) was employed to conduct all electrochemical measurements (quasi-potentiostatic and galvanostatic). Energy-efficiency measurements were performed in a monopolar electrolytic cell, which was described in our previous publication [19].

The water manometer apparatus, as described in our prior work [20], was employed to quantify the evolution time of hydrogen and oxygen (at a mole ratio of 2:1). The cathode morphology prior to and post-electrolytic process was investigated through scanning electron microscopy (SEM) analysis. The SEM images of the cathodes subjected to testing were obtained using JEOL—JSM—6610 (Jeol, Akishima, Japan) and Vega TS 5130MM (TESCAN, Brno, Czech Republic) instruments under high vacuum conditions and an accelerating voltage of 20 kV. The angle of inclination of the sample was confined between 0° and 30°, with the applied magnification being up to 3000×. The corresponding EDS spectra were captured utilizing the same instruments. XRF measurements were performed using the Thermo Scientific Niton XL3t Goldd+ XRF analyzer (Waltham, MA, USA) at room temperature.

3. Results and Discussions

To begin with, the study aimed to identify the most favorable electrolyte composition for in situ ionic activation. Table 2 and Figure 1 illustrate the voltage dependence on various Ni-Zn-Mo ion activator concentrations and applied current densities. The data indicate that combination (6) exhibited the most favorable outcomes in terms of reduced voltage and energy consumption, in comparison to the baseline electrolyte, 6 M KOH. The most significant reduction in voltage occurred at the highest current density tested, which was 100 mAcm^{-2}.

Table 2. Voltage–current density relationship at 298 K for varying concentrations of the ionic activator.

j (mA cm^{-2})	U (V)						
	6 M KOH	Ni-Zn-Mo					
		(1)	(2)	(3)	(4)	(5)	(6)
10	1.980	2.083	1.812	2.015	1.747	1.642	**1.594**
50	2.274	2.609	2.58	2.489	2.072	2.073	**1.825**
100	2.419	2.806	2.834	2.793	2.123	2.118	**1.959**

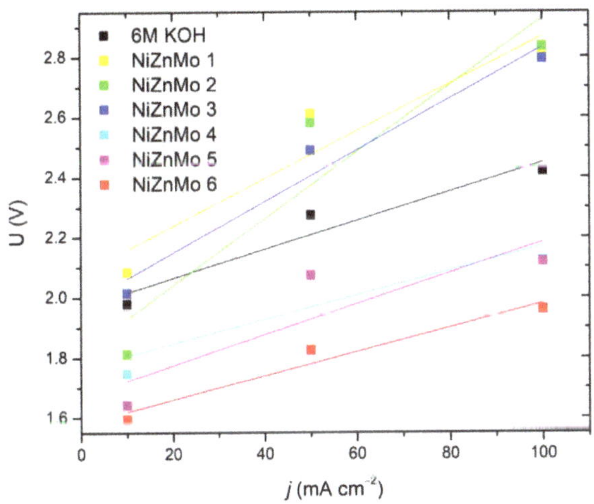

Figure 1. Voltage vs. current density of varying concentrations of the in situ added Ni-Zn-Mo ionic activators.

Detailed tests of energy consumption and reaction kinetics in alkaline electrolysis of in situ added Ni-Zn-Mo ionic activators were performed using the sixth combination of the mentioned activators (Ni-Zn-Mo (6)).

3.1. The Energy Efficiency of Water Electrolysis with In Situ Activation

The study investigated the relationship between energy consumption, temperature, and current densities in both a standard electrolyte (6 M KOH) and an electrolyte with the in situ addition of ionic activators Ni-Zn-Mo (6). The energy balance of electrolysis is represented by the energy consumed per mole of hydrogen produced, which was calculated on the basis of Formula (1):

$$Q = U \cdot I \cdot t, \quad (1)$$

where Q is the energy consumption in J mol^{-1}, U is the total voltage of electrolysis in V, i.e., the potential difference of the cathodic and anodic reactions and the voltage drop through the electrolyte, I is the total current in A, and t is the time in seconds for which 1 mol of hydrogen is developed.

The results obtained from the analysis of this system are represented by a three-dimensional diagram in Figure 2 and numerically in Table 3.

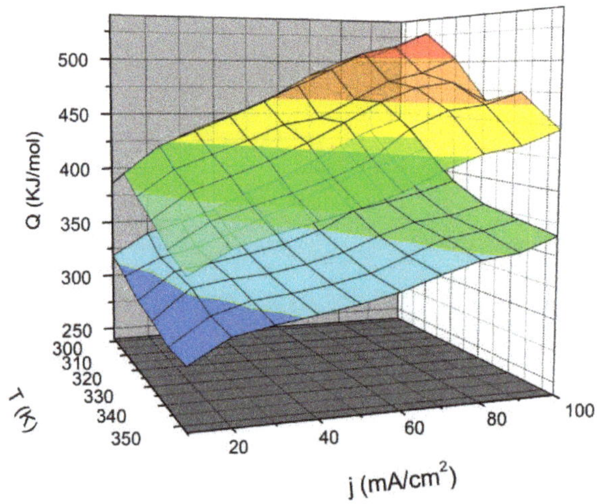

Figure 2. The relationship between energy consumption, temperature, and current density for the standard electrolyte (6 M KOH)—top surface, and with the in situ addition of the Ni-Zn-Mo ionic activator—bottom surface.

Table 3. Alkaline electrolyzer energy consumption at higher current densities and at higher temperatures.

	Q/kJmol^{-1}					
	6 M KOH			**Ni-Zn-Mo i.a.**		
	T/K			**T/K**		
***j*/mAcm^{-2}**	**303**	**323**	**333**	**303**	**323**	**333**
60	477.2	448.4	423.1	369.6	341.0	333.0
70	492.3	465.1	445.9	384.2	353.3	345.3
80	507.4	476.9	459.8	398.3	363.7	357.7
90	514.2	481.5	464.4	405.9	368.2	362.8
100	527.1	465.9	474.3	418.4	377.0	371.0

The presented diagram indicates that higher temperatures result in greater energy savings both for the standard electrolyte and the electrolyte with added ionic activators. The improvement of the electrolytic process is affected by both the elevated temperature and the addition of ionic activators. The obtained data suggest that this system has promising potential for industrial application, as even at slightly lower temperatures, the energy savings exceed 20%, significantly reducing the price of hydrogen production. The reduction in energy consumption in this system can be explained by the theory of hypo-hyper-d-inter-electronic interactions, as mentioned earlier in the introduction [21,22]. It has been concluded that electrode coatings with the appropriate composition and structure of d-metals exhibit higher activity compared to individual metals and in some cases even higher than noble metals [16,23].

Earlier works investigating the influence of in situ added $Ni(en)_3Cl_2$ and molybdates as ionic activators [24–26] show that the combination of these two d-metals significantly increases the active surface area of cathodes. Additionally, the catalytic activity of ethylene-di-amine (en), present in the electrolyte after the decomposition of the cobalt complex, plays a significant role in electrode activation. This role can be described by the Roland effect, a phenomenon observed in the electrochemical reduction of cobalt complexes, where the addition of ethylene-di-amine (en) to the electrolyte enhances the catalytic activity of cobalt toward the reduction in substrates [27,28]. This effect cleans the electrode surface by removing metal oxides [29]. Overall, the catalytic activity of ethylene-di-amine is a result of its ability to stabilize and activate the intermediate species formed during the reduction process of cobalt complexes. The formation of nickel and molybdenum deposits is promoted on the cleaned electrode surface, which in the deposition process, it undergoes the same cleaning process due to the formation of ethylene-di-amine ligands during the process. These findings support the creation of a deposit that has a highly active surface for the hydrogen evolution reaction.

3.2. The Influence of In Situ Activation on the Rate and Mechanism of the Hydrogen Evolution Reaction

The kinetics and mechanism of the hydrogen evolution reaction in systems with in situ added ionic activators were investigated by the methods of Tafel analysis and EIS spectroscopy. Polarization curves and EIS spectra were recorded for standard electrolyte solution (6 M KOH) and electrolyte with in situ added ionic activators based on Ni-Zn-Mo (6).

3.2.1. Tafel Analysis

Polarization curves for the hydrogen evolution reaction recorded for a system with a pure nickel electrode in a standard alkaline electrolyte and in an alkaline electrolyte with added ion activators are shown in Figure 3. It is observed that in situ activation results in a shift of the electrode potential of a significant 200 mV toward more positive values in a wide range of potentials. The right part of Figure 3 shows the influence of temperature on the appearance of the polarization curve for in situ activated nickel electrodes.

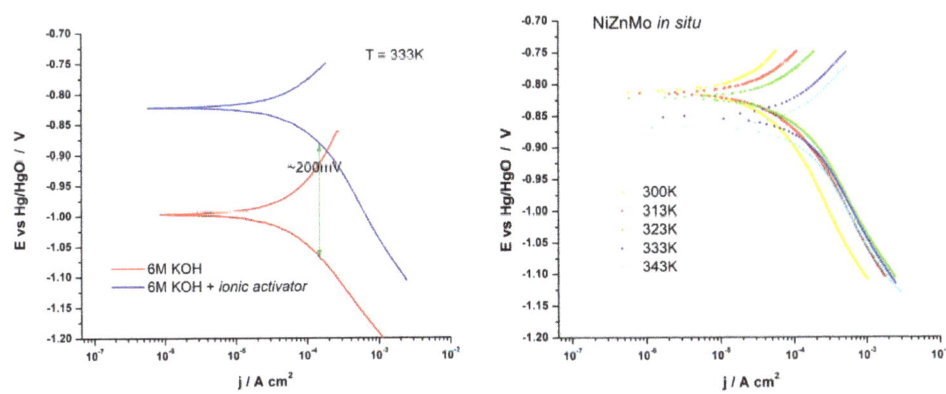

Figure 3. Polarization curves for HER in 6 M KOH with and without ionic activators (**left**) and polarization curves for HER with a Ni-Zn-Mo ionic activator added in situ in 6M KOH at temperatures from 300 to 343 K (**right**).

The basic kinetic parameters, j_0 and b, were determined by analyzing the linear part of the semilogarithmic polarization curves and presented in Table 4. The values of the Tafel slope of all recorded polarization curves are around -120 mVdec^{-1} and indicate that charge transfer (Volmer) is a slow and decisive step in the hydrogen evolution reaction. Based on the approximately equal slopes of the polarization curves both in the case of the standard electrolyte and the addition of the ionic activator, it can be concluded that in this case, the addition of Ni-Zn-Mo did not change the decisive step of the reaction, which indicates that the addition of ionic activators did not affect the change in the reaction mechanism of hydrogen evolution.

Table 4. Kinetic parameters for Ni electrodes in 6M KOH and upon addition of Ni-Zn-Mo activator.

T/K	-*b*/V dec^{-1}		j_0/A cm^{-2}	
	i.a.	6 M KOH	*i.a.*	6 M KOH
300	0.131	0.131	20.6×10^{-6}	23.95×10^{-6}
313	0.129	0.128	36.48×10^{-6}	30.38×10^{-6}
323	0.119	0.126	46.71×10^{-6}	39.66×10^{-6}
333	0.114	0.112	64.17×10^{-6}	58.16×10^{-6}
343	0.115	0.095	69.04×10^{-6}	63.15×10^{-6}

The exchange current density increases with temperature, which is the expected kinetic behavior for the observed reaction. The coating based on Ni-Zn-Mo electrochemically deposited at a temperature of 343 K, during the electrolytic process, shows the lowest values of overvoltage in the tested range of current densities and the highest value of the exchange current density, which confirms that this coating is catalytically much more active than the pure Ni electrode.

The mechanism of improving the kinetics of the electrode reaction by adding the Ni-Zn-Mo activator in situ can be described by the formation of a more developed electrode surface with a much larger number of active centers for hydrogen adsorption.

3.2.2. Electrochemical Impedance Spectroscopy

A better insight into the mechanism of the electrochemical reaction of hydrogen evolution was made possible by recording electrochemical impedance spectra for a pure nickel electrode in a standard electrolyte solution (6 M KOH) and for an electrolyte with an added Ni-Zn-Mo ionic activator. EIS spectra were recorded at overvoltages from 20 to 200 mV at temperatures from 298 to 343 K. Before each measurement, the electrodes were conditioned at a voltage of 1.3 V with respect to the SCE reference electrode for 15 s. Figure 4 shows the impedance diagrams (Nyquist diagrams) of the nickel working electrode in the electrolyte with an in situ added ionic activator.

The obtained data were fitted according to the Randels equivalent circuit model in which the capacity of the double electric layer was replaced by a constant phase element (CPE) and shown in Table 5.

The results presented in this study demonstrate that an increase in the applied overvoltage leads to a significant change in the impedance response and a decrease in the charge transfer resistance, R_{ct}, thereby indicating an enhancement in the kinetics of the hydrogen evolution reaction. At all tested temperatures, a decrease in the value of the capacity of the double electric layer, C_{dl}, can be observed with an increase in overvoltage. This can be attributed to the occlusion of the pores caused by the intense hydrogen bubble generation at higher overvoltages. At the same overvoltages, there is a noticeable increase in the value of the double electric layer capacity with increasing temperature. This is expected, considering that at higher temperatures, the resistance to charge transfer decreases. The influence of the surface on the catalytic activity can be determined by determining the roughness factor σ, which is directly related to the capacity of the electric double layer, C_{dl}. After the electrolytic process and deposition of the coating based on Ni-Zn-Mo, the surface of the electrode is no longer smooth, so an increase in roughness is expected. When C_{dl} exhibits purely capacitive behavior, the surface roughness can be estimated by assuming a certain value for a unit surface. In the case of metals, a typical value of C_{dl} is around 20 mF cm^{-2} [30]. The roughness factors (σ) were then calculated by dividing the C_{dl} value of the characterized electrode (presented in the Table 5) by that of the reference metal electrode. The obtained roughness factor (σ) values for the Ni-Zn-Mo-based ionic activators (i.a.) were found to be consistently high across all overpotentials and temperatures in comparison to pure nickel electrodes [31]. These results indicate that the improved electrocatalytic activity observed for the hydrogen evolution reaction (HER) using Ni-Zn-Mo-based i.a. is partially attributed to the increased active surface area.

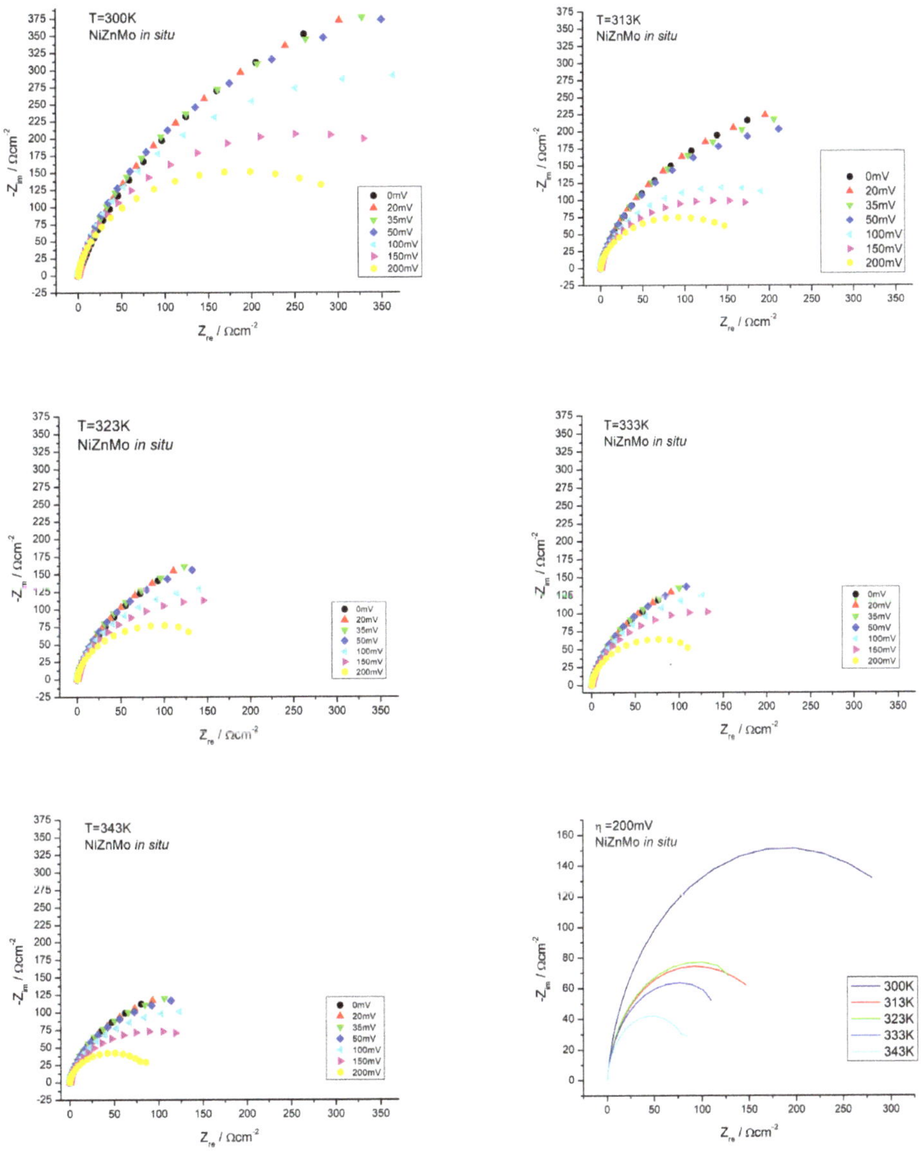

Figure 4. Nyquist plots of impedance spectra of the EIS in situ activation procedure with ionic activators based on Ni-Zn-Mo at various overvoltages and temperatures.

Table 5. Parameters obtained by fitting experimental EIS spectra recorded at different values of overvoltage and temperature for tested Ni-Zn-Mo coatings.

T (K)	$-\eta$ (mV)	R_e (Ω cm^2)	R_{ct} (Ω cm^2)	C_{dl} (mF cm^{-2})	τ/s	σ
	20	0.305	990.8	2.830	2.804	141
	35	0.338	909.1	2.429	2.208	121
	50	0.340	863.0	2.215	1.911	110
300	100	0.342	646.1	1.980	1.279	99
	150	0.342	463.0	1.834	0.849	91
	200	0.342	344.1	1.791	0.616	89
	250	0.305	990.8	2.830	2.804	141
	20	0.238	510.9	3.904	1.994	195
	35	0.237	480.5	3.638	1.748	181
	50	0.237	439.6	3.478	1.528	173
313	100	0.238	266.3	3.290	0.876	164
	150	0.238	222.2	3.351	0.744	167
	200	0.238	169.5	3.187	0.540	159
	250	0.238	510.9	3.904	1.994	195
	20	0.222	476.7	8.181	3.900	409
	35	0.224	460.3	7.240	3.332	362
	50	0.224	419.3	6.739	2.825	336
323	100	0.226	306.8	6.142	1.884	307
	150	0.228	252.7	5.467	1.381	273
	200	0.228	174.9	4.874	0.852	243
	250	0.222	476.7	8.181	3.900	409
	20	0.212	392.9	9.922	3.898	496
	35	0.214	381.3	8.745	3.334	437
333	50	0.215	362.8	7.926	2.875	396
	100	0.213	280.6	6.528	1.831	326
	150	0.209	223.1	5.570	1.242	278
	250	0.209	140.9	5.016	0.,706	250
	20	0.215	301.2	9.148	2.755	457
	35	0.212	287.1	7.762	2.228	388
343	50	0.212	267.6	6.999	1.873	349
	100	0.209	215.6	6.060	1.306	303
	150	0.205	161.9	5.365	0.868	268
	200	0.203	93.9	4.777	0.448	238

3.2.3. Elemental Composition and Morphological Characteristics

The elemental composition and morphological characteristics of the nickel cathode before and after the electrolytic process with in situ added activators based on Ni-Zn-Mo were investigated using scanning electron microscopy (SEM) and XRF analysis. The obtained results are shown in Figure 5 and in Table 6.

Presented SEM micrographs show that there is a deposition of spherical structures of nanometer dimensions evenly distributed over the entire surface of the cathode. The surface of the cathode formed by the deposition of active species after the electrolytic process is much more developed compared to the surface of a pure nickel electrode. As a result, the number of active centers for the adsorption of intermediates of the hydrogen evolution reaction is significantly increased. This confirms the conclusions reached on the basis of electrochemical measurements.

Table 6. Elemental composition of Ni-Zn-Mo coating recorded by EDS and XRF analysis.

Element	Ni	Zn	Mo	C	O
EDS	23.62%	1.21%	0.19%	32.86%	40.83%
XRF	98.1%	1.4%	0.2%	/	/

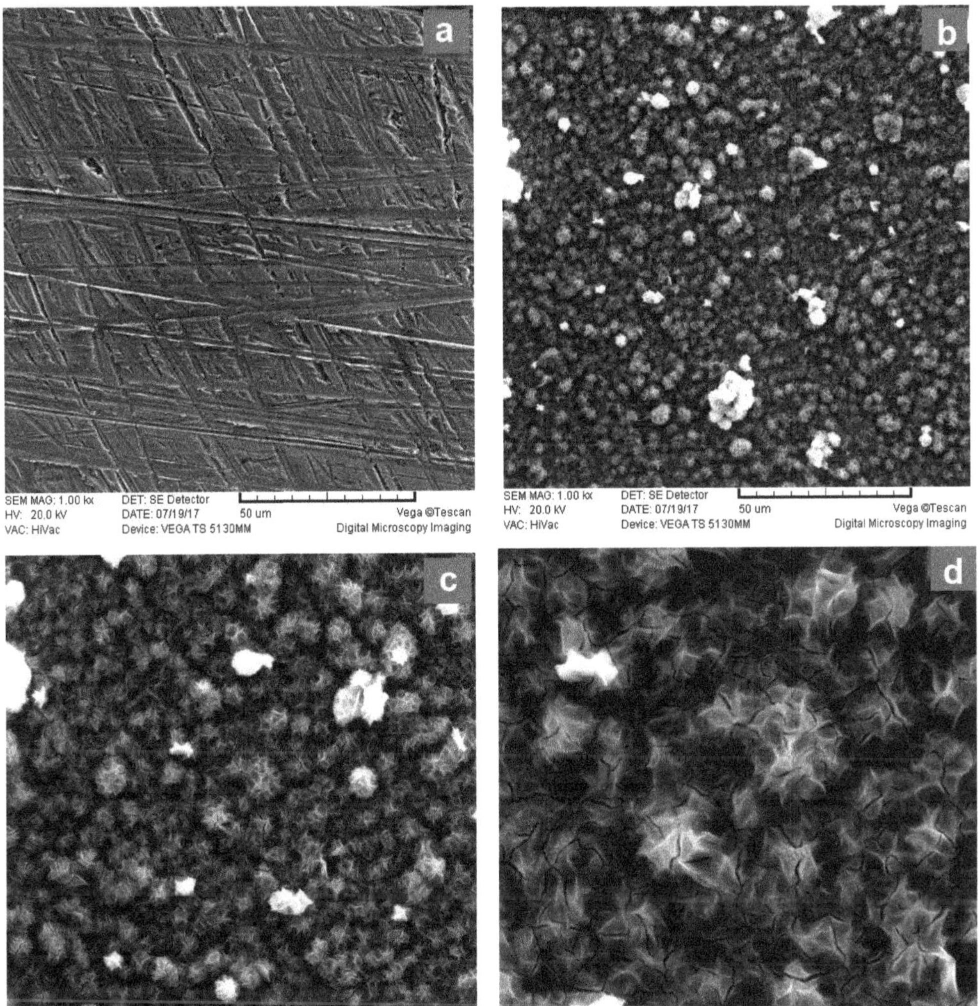

Figure 5. SEM images of: pure nickel cathode (**a**); after electrolytic process with Ni-Zn-Mo ionic activator at 1000× (**b**), 2000× (**c**) and 5000× (**d**) magnification.

The elemental composition of the resulting deposit, determined by the XRF analysis method, shows that the deposited particles are composed mostly of Ni. Zn and Mo are present in a much smaller percentage. The coating composition obtained by EDS analysis shows good agreement with the XRF analysis except in the case of nickel. This discrepancy can be attributed to the difference in the recording depth of the composite by these two methods and to the fact that the EDS analysis can see both the oxygen content resulting from exposure of the electrodes to air and the carbon content originating from the sample carrier in the SEM chamber. Given that nickel and zinc in the electrolyte are in cationic form, and molybdenum is in anionic form, it can be concluded that the induced co-deposition of these three metals occurred but to a lesser extent. Organic compounds like tris(ethylenediamine)nickel(II) chloride or dissolved gases in the electrolyte can undergo oxidation or reduction reactions, leading to the deposition of C or O onto the electrode surface. The presence of C species can be attributed to the chemical reaction involving tris(ethylenediamine)nickel(II) chloride. Furthermore, exposure of the electrode to atmospheric oxygen, in combination with the presence of sodium molybdate and tris(ethylenediamine)nickel(II) chloride in the electrolyte, can contribute to the oxidation process and explain the presence of O species. It is important to emphasize that the presence of C and O in the coating does not necessarily imply a detrimental effect on its performance.

4. Conclusions

The research findings demonstrate the superior catalytic activity of the investigated materials for the hydrogen evolution reaction compared to commercially available options. The utilization of in situ ionic activators composed of Ni-Zn-Mo can result in a notable decrease in energy consumption of up to 20% compared to a system utilizing pure nickel electrodes and a standard electrolyte solution. The observed decrease in energy consumption is pronounced at higher temperatures, whereas increasing current density is linked to Increased energy consumption. The mechanism behind the enhanced energy efficiency with the addition of ionic activators can be attributed to several interrelated effects that are involved in the overall process of in situ activation of alkaline electrolysis. One such effect is the catalytic effect, which is a synergistic consequence of the combination of transition metals. Additionally, the surface effect, which is a result of the simultaneous co-deposition of the catalytic coating and hydrogen evolution, is a critical factor. This process leads to the formation of a highly developed surface coating with numerous active centers for hydrogen adsorption. The contribution of the ethylene-di-amine ligand cannot be disregarded, as its impact is comparable to that of ethylene-di-amine tetra acetic acid (EDTA), which removes oxide layers from the cathode surface, thereby facilitating better surface development and metal deposition.

In the investigation of electrode kinetics, it was observed that the electrode activity exhibited a consistent trend of improvement, as was also noted in the energy consumption measurement experiments. Upon employing in situ activation across all tested current ranges, a marked reduction in overvoltage for the hydrogen evolution reaction was observed when compared to non-activated pure nickel electrodes operating in a standard electrolyte solution.

The investigation into the mechanism of the hydrogen evolution reaction, which was obtained by determining the values of the Tafel slopes, indicates that the slow step of the reaction in the tested system of ionic activators Ni-Zn-Mo is charge transfer, i.e., Vollmer's reaction step.

Furthermore, during the in situ activation, in addition to the hydrogen evolution reaction, a parallel process of electrochemical deposition of the alloy of d-metals is present on the cathode surface, which is easily observed based on SEM images of the formed coatings and their EDS and XRF analysis for all tested systems.

The results presented in this study demonstrate that the application of in situ ionic activators consisting of Ni-Zn-Mo can yield a substantial reduction in energy consumption and enhance electrode kinetics. This approach shows promise in advancing alkaline electrolysis and promoting the production of green hydrogen.

Author Contributions: Conceptualization, I.P. and M.M.K.; methodology, I.P., M.M.K. and G.T.; validation, I.P., G.T. and V.N.; formal analysis, I.P. and M.S.; investigation, I.P., G.T., S.M. and P.L.; resources, M.S. and. S.M.; data curation, I.P.; writing—original draft preparation, I.P.; writing—review and editing, I.P., M.M.K. and S.M.; visualization, I.P., P.L. and M.S.; supervision, M.M.K. and V.N.; project administration, V.N.; funding acquisition, M.M.K. All authors have read and agreed to the published version of the manuscript.

Funding: This work was funded by the Ministry of Science, Technological Development and Innovation of the Republic of Serbia; grant numbers: 451-03-47/2023-01/200017 and 451-03-47/2023-01/200051.

Institutional Review Board Statement: Not applicable.

Informed Consent Statement: Not applicable.

Data Availability Statement: Not applicable.

Acknowledgments: The authors would like to thank Ministry of Science, Technological Development and Innovation of the Republic of Serbia for financial support (grant numbers 451-03-47/2023-01/200017 and 451-03-47/2023-01/200051).

Conflicts of Interest: The authors declare no conflict of interest.

References

1. Zerrahn, A.; Schill, W.-P.; Kemfert, C. On the Economics of Electrical Storage for Variable Renewable Energy Sources. *Eur. Econ. Rev.* **2018**, *108*, 259–279. [CrossRef]
2. Bockris, J. *Energy: The Solar-Hydrogen Alternative*; Architectural Press: New York, NY, USA, 1976; ISBN 0855520426.
3. Hauglustaine, D.; Paulot, F.; Collins, W.; Derwent, R.; Sand, M.; Boucher, O. Climate Benefit of a Future Hydrogen Economy. *Commun. Earth Environ.* **2022**, *3*, 295. [CrossRef]
4. Martínez-Rodríguez, A.; Abánades, A. Comparative Analysis of Energy and Exergy Performance of Hydrogen Production Methods. *Entropy* **2020**, *22*, 1286. [CrossRef] [PubMed]
5. Panigrahy, B.; Narayan, K.; Ramachandra Rao, B. Green Hydrogen Production by Water Electrolysis: A Renewable Energy Perspective. *Mater. Today Proc.* **2022**, *67*, 1310–1314. [CrossRef]
6. Shin, H.; Jang, D.; Lee, S.; Cho, H.-S.; Kim, K.-H.; Kang, S. Techno-Economic Evaluation of Green Hydrogen Production with Low-Temperature Water Electrolysis Technologies Directly Coupled with Renewable Power Sources. *Energy Convers. Manag.* **2023**, *286*, 117083. [CrossRef]
7. Wang, S.; Lu, A.; Zhong, C.-J. Hydrogen Production from Water Electrolysis: Role of Catalysts. *Nano Converg.* **2021**, *8*, 4. [CrossRef]
8. Roger, I.; Shipman, M.A.; Symes, M.D. Earth-Abundant Catalysts for Electrochemical and Photoelectrochemical Water Splitting. *Nat. Rev. Chem.* **2017**, *1*, 0003. [CrossRef]
9. Brewer, L. *Chemical Bonding Theory Applied to Metals*; ASM International: New York, NY, USA, 1988.
10. Dubouis, N.; Grimaud, A. The Hydrogen Evolution Reaction: From Material to Interfacial Descriptors. *Chem. Sci.* **2019**, *10*, 9165–9181. [CrossRef]
11. Krstajic, N.; Jovic, V.; Gajickrstajic, L.; Jovic, B.; Antozzi, A.; Martelli, G. Electrodeposition of Ni–Mo Alloy Coatings and Their Characterization as Cathodes for Hydrogen Evolution in Sodium Hydroxide Solution. *Int. J. Hydrogen Energy* **2008**, *33*, 3676–3687. [CrossRef]
12. Tavares, A.C.; Trasatti, S. Ni + RuO 2 Co-Deposited Electrodes for Hydrogen Evolution. *Electrochim. Acta* **2000**, *45*, 4195–4202. [CrossRef]
13. Obradović, M.D.; Stevanović, R.M.; Despić, A.R. Electrochemical Deposition of Ni–W Alloys from Ammonia–Citrate Electrolyte. *J. Electroanal. Chem.* **2003**, *552*, 185–196. [CrossRef]
14. Jakšić, J.M.; Krstajić, N.V.; Grgur, B.N.; Jakšić, M.M. Hydridic and Electrocatalytic Properties of Hypo-Hyper-d-Electronic Combinations of Transition Metal Intermetallic Phases. *Int. J. Hydrogen Energy* **1998**, *23*, 667–681. [CrossRef]
15. Nikolic, V.M.; Maslovara, S.L.; Tasic, G.S.; Brdaric, T.P.; Lausevic, P.Z.; Radak, B.B.; Marceta Kaninski, M.P. Kinetics of Hydrogen Evolution Reaction in Alkaline Electrolysis on a Ni Cathode in the Presence of Ni–Co–Mo Based Ionic Activators. *Appl. Catal. B Environ.* **2015**, *179*, 88–94. [CrossRef]
16. Jaksic, M. Brewer Intermetallic Phases as Synergetic Electrocatalysts for Hydrogen Evolution. *Mater. Chem. Phys.* **1989**, *22*, 1–26. [CrossRef]
17. Miulovic, S.M.; Maslovara, S.L.; Perovic, I.M.; Nikolic, V.M.; Marceta Kaninski, M.P. Electrocatalytic Activity of ZnCoMo Based Ionic Activators for Alkaline Hydrogen Evolution-Part II. *Appl. Catal. A Gen.* **2013**, *451*, 220–226. [CrossRef]
18. Maslovara, S.L.; Aničijević, D.V.; Kijevcanin, M.L.; Radovic, I.R.; Nikolic, V.M.; Lausevic, P.Z.; Marceta Kaninski, M.P. Improved HER Activity of Ni and Stainless Steel Electrodes Activated by NiCoMo Ionic Activator—A Combined DFT and Experimental Study. *Int. J. Hydrogen Energy* **2017**, *42*, 5072–5082. [CrossRef]
19. Perovic, I.M.; Acimovic, D.D.; Tasić, G.S.; Karic, S.D.; Lausevic, P.Z.; Marčeta Kaninski, M.P.; Nikolić, V.M. Efficient Hydrogen Production Using Ternary Ni-Cu-Mo Ionic Activator. *Int. J. Hydrogen Energy* **2015**, *40*, 6270–6275. [CrossRef]
20. Nikolic, V.M.; Tasic, G.S.; Maksic, A.D.; Saponjic, D.P.; Miulovic, S.M.; Marceta Kaninski, M.P. Raising Efficiency of Hydrogen Generation from Alkaline Water Electrolysis—Energy Saving. *Int. J. Hydrogen Energy* **2010**, *35*, 12369–12373. [CrossRef]
21. Jaksic, J.; Lacnjevac, C.; Krstajic, N.; Jaksic, M. Interactive Supported Electrocatalysts and Spillover Effect in Electrocatalysis for Hydrogen and Oxygen Electrode Reactions. *Chem. Ind. Chem. Eng. Q.* **2008**, *14*, 119–136. [CrossRef]

22. Neophytides, S.G.; Zafeiratos, S.; Papakonstantinou, G.D.; Jaksic, J.M.; Paloukis, F.E.; Jaksic, M.M. Extended Brewer Hypo–Hyper-d-Interionic Bonding Theory—I. Theoretical Considerations and Examples for Its Experimental Confirmation. *Int. J. Hydrogen Energy* **2005**, *30*, 131–147. [CrossRef]
23. Jaksic, M.M. Hypo—Hyper-d-Electronic Interactive Nature of Synergism in Catalysis and Electrocatalysis for Hydrogen Reactions. *Electrochim. Acta* **2000**, *45*, 4085–4099. [CrossRef]
24. Safizadeh, F.; Ghali, E.; Houlachi, G. Electrocatalysis Developments for Hydrogen Evolution Reaction in Alkaline Solutions—A Review. *Int. J. Hydrogen Energy* **2015**, *40*, 256–274. [CrossRef]
25. Zhang, S.; Zhang, X.; Rui, Y.; Wang, R.; Li, X. Recent Advances in Non-Precious Metal Electrocatalysts for PH-Universal Hydrogen Evolution Reaction. *Green Energy Environ.* **2021**, *6*, 458–478. [CrossRef]
26. Bau, J.A.; Kozlov, S.M.; Azofra, L.M.; Ould-Chikh, S.; Emwas, A.-H.; Idriss, H.; Cavallo, L.; Takanabe, K. Role of Oxidized Mo Species on the Active Surface of Ni–Mo Electrocatalysts for Hydrogen Evolution under Alkaline Conditions. *ACS Catal.* **2020**, *10*, 12858–12866. [CrossRef]
27. Rowland, P.R. Electrolytic Separation Factor of Protium and Deuterium. *Nature* **1968**, *218*, 945–946. [CrossRef]
28. Rowland, P.R. The Electrolytic H/D Separation Factor-Agreement of the Experiment with the Hypothetical Maximum Value. *J. Electroanal. Chem. Interfacial Electrochem.* **1971**, *32*, 109–124. [CrossRef]
29. Paoletti, P. Formation of Metal Complexes with Ethylenediamine: A Critical Survey of Equilibrium Constants, Enthalpy and Entropy Values. *Pure Appl. Chem.* **1984**, *56*, 491–522. [CrossRef]
30. Stolten, D.; Emonts, B. *Hydrogen Science and Engineering: Materials, Processes, Systems and Technology*; John Wiley & Sons: Hoboken, NJ, USA, 2016; ISBN 3527332383.
31. Kibria, M.F.; Mridha, M.S. Electrochemical Studies of the Nickel Electrode for the Oxygen Evolution Reaction. *Int. J. Hydrogen Energy* **1996**, *21*, 179–182. [CrossRef]

materials MDPI

Article

Achieving High Activity and Long-Term Stability towards Oxygen Evolution in Acid by Phase Coupling between CeO_2-Ir

Jianren Kuang [1], Zhi Li [1], Weiqiang Li [2], Changdong Chen [3], Ming La [3,*] and Yajuan Hao [2,*]

1 College of Environment and Energy, South China University of Technology, Guangzhou 510006, China; kjr_be_a_man1998@163.com (J.K.); m19927657431@163.com (Z.L.)
2 College of Electric and Information Engineering, Pingdingshan University, Pingdingshan 467000, China; 5001@pdsu.edu.cn
3 College of Chemistry and Environmental Engineering, Pingdingshan University, Pingdingshan 467000, China; lmccd5613@163.com
* Correspondence: laming82@126.com (M.L.); yjhao@pdsu.edu.cn (Y.H.)

Abstract: The development of efficient and stable catalysts with high mass activity is crucial for acidic oxygen evolution reaction (OER). In this study, CeO_2-Ir heterojunctions supported on carbon nanotubes (CeO_2-Ir/CNTs) are synthesized using a solvothermal method based on the heterostructure strategy. CeO_2-Ir/CNTs demonstrate remarkable effectiveness as catalysts for acidic OER, achieving 10.0 mA cm^{-2} at a low overpotential of only 262.9 mV and maintaining stability over 60.0 h. Notably, despite using an Ir dosage 15.3 times lower than that of c-IrO_2, CeO_2-Ir/CNTs exhibit a very high mass activity (2542.3 A $g_{Ir}{}^{-1}$@1.53 V), which is 58.8 times higher than that of c-IrO_2. When applied to acidic water electrolyzes, CeO_2-Ir/CNTs display a prosperous potential for application as anodic catalysts. X-ray photoelectron spectrometer (XPS) analysis reveals that the chemical environment of Ir nanoparticles (NP) can be effectively modulated through coupling with CeO_2. This modulation is believed to be the key factor contributing to the excellent OER catalytic activity and stability observed in CeO_2-Ir/CNTs.

Keywords: CeO_2-Ir heterojunction; electrocatalysis; acidic oxygen evolution reaction; high activity and long-term stability

1. Introduction

A significant amount of renewable energy (such as wind energy, tidal energy, solar energy, etc.) is wasted due to ineffective utilization. Utilizing energy storage and conversion devices to convert renewable energy into other forms of stored energy represents an effective approach for addressing the aforementioned issues [1,2]. A proton exchange membrane water electrolyzer (PEMWE) offers the capability to convert renewable energy into chemical energy in hydrogen, which possesses sustainable, clean, and efficient characteristics [3,4]. PEMWE comprises two crucial half-reactions: cathodic hydrogen evolution reaction (HER) and anodic oxygen evolution reaction (OER). OER involves a four electron–proton transfer process that includes complex intermediates' conversions and necessitates high kinetic costs [5,6]. Most catalysts reported thus far encounter a common challenge: low catalytic activity and poor stability in acidic OER [7,8]. In view of the comprehensive consideration of catalytic activity and stability, Ir-based materials are widely used as catalysts in acidic OER. However, these catalysts currently face challenges such as scarcity, high cost, insufficient catalytic activity, and poor stability, hence they still have a long way to go before practical application can be achieved [9,10]. Consequently, exploring low cost Ir-based catalysts with high activity and durability has become a challenge at present.

It has been reported that metallic Ir catalysts with nanorods [11], nanosheets [12], and other structures exhibit exceptional catalytic activity and stability in acidic OER [13]. Significant efforts have been devoted to minimizing the use dosage of Ir without compromising the catalytic activity. Downsizing Ir to the nanometer scale can enhance the exposure of active sites, thereby improving catalytic activity [14]. Furthermore, the construction of heterojunctions could induce lattice strain within the active phase and allow for the regulation of charge distribution between the two phases, resulting in a regulation of their catalytic activity [15,16]. Consequently, the electronic structure of Ir can be modulated through the construction of heterojunctions with other phases, indirectly influencing the adsorption and desorption of intermediates [16,17]. Appropriate adsorption/desorption facilitates conversion processes of intermediates, leading to enhanced reaction rates and accelerated kinetics [18,19]. The construction of a heterojunction between Ir and metal oxides is a common strategy for enhancing both catalytic activity and stability in Ir-based catalysts. Lee et al. achieved improved OER performance by combining Ir with MoO_3 to construct highly electron-deficient Ir NPs [16]. Xing and colleagues uniformly deposited an ultra-fine (~1.0 nm) layer of Ir metal onto the surface of niobium oxide (Nb_2O_{5-x}), which exhibited a high concentration of oxygen vacancies [20]. During OER, dynamic migration of oxygen takes place at the two-phase interface between Ir and Nb_2O_{5-x}, effectively preventing excessive oxidation of Ir to high-valent species and subsequent deactivation. Although extensive research has focused on developing heterogeneous catalysts based on transition metal oxides combined with iridium (Ir/MO_x), investigations into heterojunctions formed by rare earth metal oxides and iridium remain scarce. Cerium oxide is widely used as a catalyst and cocatalyst for many oxidation reactions due to its excellent redox properties [21–24]. For example, Im and colleagues improved the electrocatalytic activity of OER by inducing site-selective crystal disorder through the doping of Ce^{3+} ions into the MIL-88B(Ni) framework [25]. The aforementioned statements have prompted us to redirect our research focus towards CeO_2 in order to investigate the electron transfer between CeO_2 and Ir, as well as its impact on both the activity and the stability of OER. Moreover, carbon materials are widely acknowledged as ideal supports for electrocatalysis owing to their remarkable specific surface area and excellent electrical conductivity [26–28]. As an exemplary carrier, CNTs offer abundant nucleation sites and conductive pathways for catalytic active materials, thereby significantly enhancing the catalytic activity of catalysts while reducing their required quantity [29,30].

In this study, we propose a strategy for the construction of heterojunctions and carbon combination to enhance catalytic performance while reducing the consumption of Ir. This strategy involves modulating the electronic structure of Ir and improving the electron conduction of the catalyst. To achieve this, we engineer electron-deficient Ir NPs through the construction of heterojunctions with CeO_2, complemented by the incorporation of CNTs to facilitate efficient charge conduction and improve the utilization of Ir. As a result, CeO_2-Ir/CNTs exhibit impressive durability and catalytic activity, operating continuously at 10.0 mA cm^{-2} for over 60.0 h with only 262.9 mV overpotential. These findings underscore the excellent catalytic performance and long-term robustness of CeO_2-Ir/CNTs when employed in acidic water electrolyzers.

2. Materials and Methods

2.1. Chemicals

Multi-walled carbon nanotubes (CNTs, >97.0%) were bought from Shenzhen Nanotech Port Co., LTD, Shenzhen, China. Iridium (III) chloride ($IrCl_3$, with Ir $\geq$ 62.0%) and cerous chloride ($CeCl_3$, 99.9%) were purchased from Macklin. Carbon papers (CPs) were obtained from Shanghai Hesen Electric Co., Ltd., Shanghai, China; 20 wt.% Pt/C was from Aladdin. Concentrated sulfuric acid (H_2SO_4, 95.0~98.0%) and nitric acid (HNO_3, 65.0~68.0%) were obtained from the Guangzhou chemical reagent factory. Nafion D520 (5.0% in isopropanol) was purchased from DuPont. The commercial iridium (IV) oxide (c-IrO_2, metal basis, Ir $\geq$ 84.5%) was from Energy Chemical. All the chemicals were used as obtained.

2.2. Synthesis of $xCeO_2$-Ir/CNTs and CeO_2/CNTs

CeO_2-Ir/CNTs ($xCeO_2$-Ir/CNTs, x = 1) were synthesized via a solvothermal method. Specifically, 4.0 mL of 5.0 mM $IrCl_3$, 4.0 mL of 5.0 mM $CeCl_3$, and 3.0 mL of 5.0 g/mL acid-treated CNTs [30] were sequentially added dropwise to 15.0 mL ethanol. Water was added to regulate the volume of the above solution to 30.0 mL, followed by thorough agitation. The obtained mixture was transferred to a 50.0 mL Teflon autoclave and heated at 180.0 °C for 2.0 h. After cooling, the black precipitate was collected by centrifugation and alternate washing with ethanol and water for three runs, followed by vacuum drying. The $xCeO_2$-Ir/CNTs were synthesized by varying the feeding volume of $CeCl_3$, where x refers to the molar ratio of $CeCl_3/IrCl_3$, albeit keeping other parameters constant. Similarly, CeO_2/CNTs were synthesized under the same conditions but without $IrCl_3$.

2.3. Characterization

Morphologies and microstructures of samples were characterized using field-emission scanning electron microscopy (FESEM, Hitachi SU8010) and transmission electron microscopy (TEM, Talos F200X equipped with an energy dispersive X-ray energy spectrometer (EDS). An X-ray diffractometer (XRD, Bruker D8-Advance) with Cu Kα radiation was applied to collect XRD patterns in a scanning range of 10.0 to 90.0° at a rate of 10.0°/min. X-ray photoelectron spectra (XPS) was recorded with an X-ray photoelectron spectrometer (XPS, Thermo ESCA-LAB 250XI) with monochromatic Al Kα excitation source, in which the binding energies were calibrated with C 1 s at 284.6 eV. The elemental composition analysis was conducted on an inductively coupled plasma optical emission spectrometer (ICP-OES, iCAP 7200 Duo). The thermogravimetric (TGA) curve was recorded on a NETZSCH STA 449 F5 under the air stream, starting at 30.0 °C and ramping up to 800.0 °C at a rate of 10.0 °C/min.

2.4. Electrochemical Measurements

All electrochemical tests were conducted using a three-electrode system with an electrochemical workstation (CHI 660e, Chenghua, Shanghai). A Ag/AgCl electrode and a Pt wire were utilized as the reference and counter electrodes, respectively. The electrolyte used was a 0.5 M H_2SO_4 solution with a pH of 0.32 (Figure S1) at room temperature, consistent with values reported in the existing literature [31,32]. The catalyst-loaded glassy carbon electrode (GCE) was prepared by drop-casting the catalyst ink onto the GCE, and utilized as the working electrode after natural drying. The catalyst ink was prepared by dispersing 4.0 mg catalyst in 1.0 mL of isopropanol/water solution (V/V = 7/3) with 10.0 µL Nafion. The mass loading of the electrocatalysts was estimated to be ~0.20 mg cm^{-2}. The linear sweep voltammograms (LSVs) were recorded with a scan rate of 5.0 mV s^{-1} at 1600 rpm to evaluate the catalytic performance of the catalysts. The LSVs reported in this work were 95% iR corrected [33–35] unless otherwise stated, and the potentials reported were calibrated with the reversible hydrogen electrode (RHE) using the equation $E(V \text{ vs. } RHE) = E(Ag/AgCl) + 0.059 \times pH$. The electrochemical impedance spectra (EIS) were obtained at 1.32 V in the frequency range from 100 KHz to 0.01 Hz. The cyclic voltammograms (CVs) were measured within the non-Faradic potential range at different scan rates (v). The electrochemical double layer capacitance (C_{dl}) of catalysts was calculated by plotting half the difference between the anode and cathode current densities ($\Delta j = j_{anode} - j_{cathode}$) vs. v. The electrochemically active surface areas (ECSAs) of the catalysts were caculated according to $S_{ECSA} = C_{dl} \times S/C_s$ (where S and C_s represent the geometric area of the electrode and the roughness factor with a value of 35.0 µF cm^{-2} [36], respectively). According to $j_{ECSA} = j_{geo}/S_{ECSA}$, ECSA-normalized LSVs of the catalysts were estimated (j_{geo} denotes the geometric activity of the catalysts). The chronopotentiometric curves were used to evaluate the catalysts' stability on the carbon paper (CP) with a mass loading of 2.0 mg cm^{-2}. The catalyst-loaded CP mentioned in the text (active area: 1.0×1.0 cm^2, Figure S2) was prepared using the same procedure as for GCE.

The Faradaic efficiency (FE) of CeO_2-Ir/CNTs towards OER on CP was performed in a three-electrode system and investigated using a gas chromatograph (GC, 9560, Shanghai Huaai Scientific Instrument, Shanghai, China). Electrolysis was carried out for a specific duration at a current density of 10.0 mA cm^{-2}. Following each electrolysis stage, gaseous samples were drawn from the headspace using a gas-tight syringe (with multiple extractions for averaging), and subsequently analyzed by a GC. Herein, FE of O_2 was calculated according to equation $FE(O_2) = V(O_2) \times 4 \times F/(V_m \times i \times t) \times 100\%$. $V(O_2)$ is the generated volume of O_2, F is the Faraday constant of 96,485.3 C/mol, V_m denotes the molar volume of the gas, i is the recorded current, and t is the time spent in electrolysis.

2.5. Evaluation of Water Electrolyzers

Practical usability of the catalyst was appraised by assembling a water electrolyzer using Pt/C and the CeO_2-Ir/CNTs as the cathode and anode catalysts, respectively. Both Pt/C and CeO_2-Ir/CNTs were loaded on the CPs. Their mass loadings were both controlled to be 2.0 mg cm^{-2}. The electrochemical tests were carried out in 0.5 M H_2SO_4. To give a comparison, water electrolyzers with Pt/C and c-IrO_2 were also prepared.

3. Results and Discussion

The synthetic scheme for CeO_2-Ir/CNTs is presented in Figure 1a. Briefly, CeO_2-Ir/CNTs were synthesized by maintaining a solution containing CNTs, $IrCl_3$, and $CeCl_3$ at 180.0 °C for 2.0 h. FESEM was carried out to describe the morphology of CeO_2-Ir/CNTs, as depicted in Figure S3a. The surface of each CNT appeared rough and was densely attached with ultra-fine NPs. The morphologies of Ir/CNTs and CeO_2/CNTs were similar to that of CeO_2-Ir/CNTs (Figure S3b,c), indicating that the addition of $CeCl_3$ or $IrCl_3$ had negligible effects on the samples' morphologies. The micro-structure of CeO_2-Ir/CNTs was further investigated by TEM. As depicted in Figure 1b, CeO_2 and Ir NPs were well-dispersed on the surface of CNT. Size distribution analysis revealed their average diameter was approximately 2.15 nm (Figure 1b). HRTEM, in Figure 1c, demonstrated multiple paired nanoparticles (marked with yellow calabash), where each nanoparticle displayed clear lattice fringes, indicating their excellent crystallinities. The FFT images in Figure 1d,e clearly illustrate that the paired nanoparticles showed different lattice fringe spacings, that is, 0.221 and 0.192 nm, corresponding well to the (111) plane of Ir (PDF No. 87-0715) and the (220) plane of CeO_2 (PDF No. 89-8436), respectively. These results suggest the successful construction of CeO_2-Ir heterojunctions (Figure 1c). Furthermore, energy-dispersive X-ray spectroscopy (EDS) mapping demonstrated uniform distribution along the contour of CeO_2-Ir/CNTs for C, O, Ir, and Ce elements (Figure 1f). XRD patterns of CeO_2-Ir/CNTs and CeO_2/CNTs are provided in Figure 2a and Figure S4, respectively. The diffraction peaks of Ir (111), CeO_2 (111), and (200) of CNTs were observed for CeO_2-Ir/CNTs, fully consistent with HRTEM observations. Meanwhile, only those of CeO_2 and CNTs were visible for CeO_2/CNTs. The composition of xCeO_2-Ir/CNTs was investigated by ICP-OES and the results are summarized in Table S1. It was clearly seen that the Ce/Ir molar ratio of xCeO_2-Ir/CNTs samples was almost consistent with the feeding ratio of $CeCl_3$/$IrCl_3$. The additional amount of Ce did not alter the structure of xCeO_2-Ir/CNTs. As depicted in Figure S5a, irrespective of the CeO_2 content in xCeO_2-Ir/CNTs, their XRD diffraction peaks consistently exhibited a combination pattern comprising CeO_2, Ir, and CNTs; only the peak of CeO_2 (111) grew higher with the increasing CeO_2 content. Thermogravimetric analysis (TGA) was further applied to calculate the weight percentage of the components within CeO_2-Ir/CNTs. According to Figure 2b, a weight loss of ~88.5 wt.% was obviously observed at 800.0 °C in air, where the CeO_2-Ir/CNTs was transformed into CeO_2 and IrO_2. Consequently, by integrating the ICP-OES and TGA analysis, it was inferred that the weight percentage of Ir in CeO_2-Ir/CNTs was estimated to be 5.6 wt.%. EDS analysis further confirmed a 1:1 molar ratio of Ce/Ir and an Ir content of 5.6 wt.% in CeO_2-Ir/CNTs, as shown in Table S2.

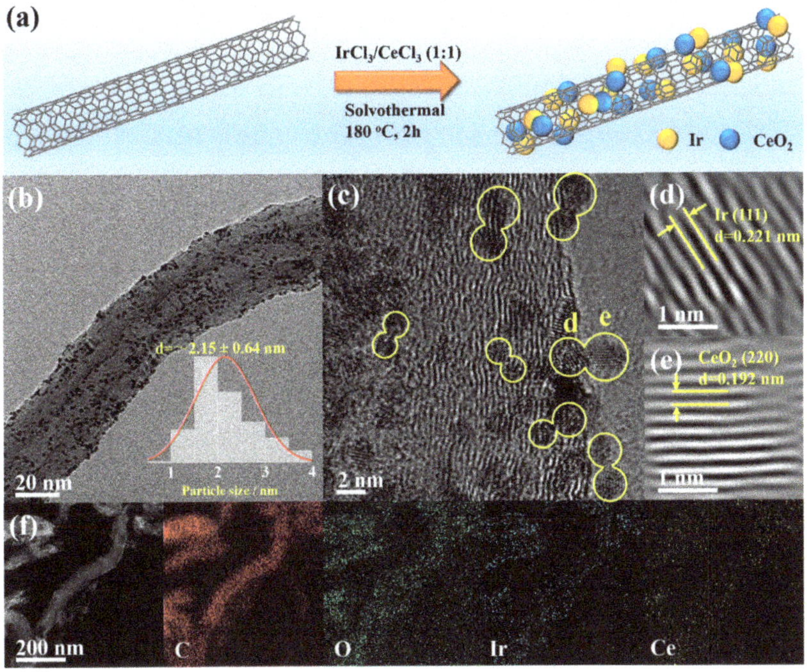

Figure 1. (**a**) Brief illustration of the synthetic procedure for CeO_2-Ir/CNTs. (**b**) TEM, (**c**) HRTEM, and (**f**) EDS mapping images of CeO_2-Ir/CNTs. (**d**,**e**) represent Ir and CeO_2 nanoparticle lattice fringes from (**c**), respectively.

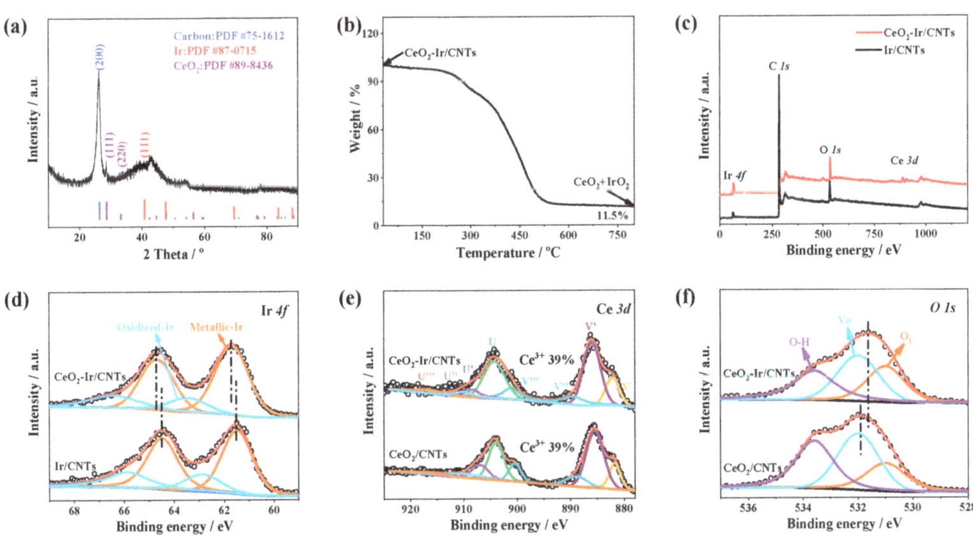

Figure 2. (**a**) XRD pattern and (**b**) TGA curve of CeO_2-Ir/CNTs. (**c**) Survey XPS spectra. XPS spectra of (**d**) Ir 4f, (**e**) Ce 3d, (**f**) O1s.

XPS measurement was used to reveal the valance state of CeO_2-Ir/CNTs. As illustrated in Figure 2c, the survey XPS spectra confirmed the presence of Ce, Ir, O, and C in the sample of CeO_2-Ir/CNTs. The Ir 4f spectra were deconvoluted into the following two sets of peaks: metallic Ir peaks and oxidized Ir peaks. The peaks at 61.5 and 64.5 eV could be assigned to Ir $4f^{7/2}$ and Ir $4f^{5/2}$ of metallic Ir; the oxidized Ir were at 62.9 and 65.9 eV (Figure 2d) [11,37]. The XPS spectra of Ce 3d of CeO_2-Ir/CNTs are shown in Figure 2e. From Figure 2e, the Ce 3d spectrum could be decomposed into four pairs of spin orbitals (V/U, V′/U′, V″/U″, and V‴/U‴), where V and U represent the $3d^{3/2}$ and $3d^{5/2}$ states, respectively [38]. The peaks corresponding to V/U, V″/U″, and V‴/U‴ were attributed to Ce^{4+} 3d orbitals, while the peaks corresponding to V′/U′ were associated with Ce^{3+} 3d [39]. The ratio of $Ce^{3+}/Ce^{3+} + Ce^{4+}$ in CeO_2 kept constant (39%) after coupling with Ir (Figure 2e), indicating that the chemical environment of Ce did not undergo significant changes. The O 1s spectra in Figure 2f could be deconvoluted into three peaks: O_L (531.0 eV), Vo· (532.0 eV), and O-H (533.6 eV), where O_L and Vo· refer to oxygen in the lattice and oxygen vacancies, respectively. The Vo· peak further verified the existence of Vo· in the CeO_2, in line with the presence of a large amount of Ce^{3+} species. Closer examinations revealed a strong electron coupling between CeO_2 and Ir through the heterointerface in the CeO_2-Ir/CNTs. Figure 2d shows that the peaks corresponding to Ir 4f in the CeO_2-Ir/CNTs appeared at a relatively higher binding energy compared with Ir/CNTs (synthesized in the absence of Ce). Remarkably, after being coupled with Ir, the O 1s peaks of CeO_2-Ir/CNTs appeared at a relatively lower binding energy compared with CeO_2/CNTs (synthesized in the absence of Ir). These observations support the conclusion that the coupling between CeO_2 and Ir NPs results in strong electron coupling, particularly occurring between Ir and the O_L of CeO_2.

The influence of CeO_2 content on the catalytic activity of xCeO_2-Ir/CNTs was investigated. As illustrated in Figure S5b, the catalyst activity was improved, by increasing x from 0 to 1.0. However, further increasing x to 1.5 led to declined activity. These findings suggest that optimal OER activity could be achieved by maintaining the Ce/Ir molar ratio of xCeO_2-Ir/CNTs at 1/1. Furthermore, Tafel plots (Figure S5c) and EIS analysis (Figure S5d) revealed that the OER kinetics and charge transfer resistance of xCeO_2-Ir/CNTs highly depended on x, and both reached the best at x = 1.0. The electrochemical performance of the synthesized CeO_2/CNTs and the purchased c-IrO_2 were also studied as comparisons. Specifically, CeO_2-Ir/CNTs demonstrated remarkable catalytic activity in the acidic OER test. As depicted in Figure 3a, CeO_2-Ir/CNTs demonstrated remarkable catalytic activity with an overpotential of 262.9 mV at 10.0 mA cm^{-2} under the acidic OER conditions (0.5 M H_2SO_4), surpassing that of Ir/CNTs (285.9 mV), c-IrO_2 (304.1 mV), and CeO_2/CNTs (with negligible OER current). Furthermore, CeO_2-Ir/CNTs even outperformed numerous recently reported Ir-based catalysts (Table S3). The superior catalytic activity of CeO_2-Ir/CNTs was further evidenced by its accelerated OER kinetics. As illustrated in Figure 3b, CeO_2-Ir/CNTs exhibited a Tafel slope of 53.4 mV dec^{-1}, lower than that of Ir/CNTs (59.2 mV dec^{-1}) and c-IrO_2 (69.5 mV dec^{-1}). Additionally, mass activities of CeO_2-Ir/CNTs, Ir/CNTs, and c-IrO_2 are plotted in Figure 3c. Among them, CeO_2-Ir/CNTs exhibited a mass activity of 2542.3 A g_{Ir}^{-1} at 1.53 V, 1.8 and 58.8 times higher than that of Ir/CNTs and c-IrO_2, respectively. EIS was performed to elucidate the origin of the superior catalytic activity of CeO_2-Ir/CNTs. At a potential of 1.32 V, CeO_2-Ir/CNTs displayed a significantly reduced charge transfer resistance value of 18.1 Ω compared with c-IrO_2 (46.5 Ω) and other catalysts (Figure 3d). CeO_2-Ir/CNTs efficiently converted electrical power into chemical power, as evidenced by its remarkable FE of 98.2% (Figure S6). It is important to note that the FE of CeO_2-Ir/CNTs was less than 100%, which could be attributed to minor carbon corrosion during the OER [40].

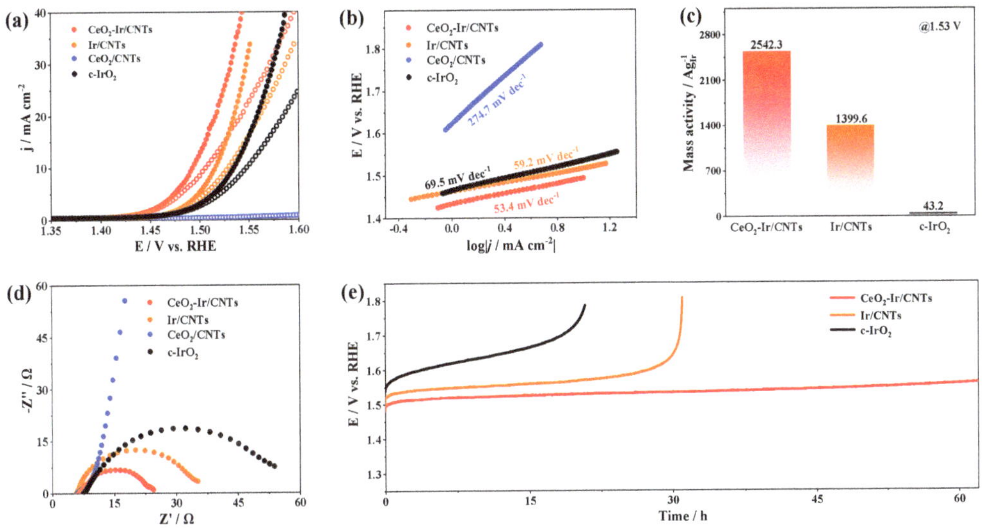

Figure 3. (**a**) LSVs (solid−dot line with iR compensation), (**b**) Tafel slopes, (**c**) mass activity, (**d**) EIS Nyquist plots, and (**e**) chronopotentiometric curves of different catalysts.

CV tests were conducted at different scan rates (v) in the non-Faradic region to calculate the C_{dl} of catalysts, as depicted in Figure S7a–e. To exclude the morphological effects caused by CeO_2 and to investigate the intrinsic catalytic activity of the catalysts, LSVs for xCeO_2-Ir/CNTs were normalized by their ECSAs that were determined by using a CV method, as depicted in Figure S7f. Notably, CeO_2-Ir/CNTs displayed a relatively larger ECSA (46.0 cm^2) compared with Ir/CNTs (42.1 cm^2), 0.5CeO_2-Ir/CNTs (45.4 cm^2), and 1.5CeO_2-Ir/CNTs (41.5 cm^2). Nevertheless, after the ECSA normalization, CeO_2-Ir/CNTs still showed a larger current density than others, indicating their superior intrinsic activity (Figure S8). Thus, the CeO_2 content had a notable influence on the intrinsic catalytic activity of xCeO_2-Ir/CNTs. Considering the XPS analysis presented in Figure 2d,f, it could be inferred that the electron interaction between O_L of CeO_2 and Ir led to the generation of electron-deficient Ir atoms. DFT calculations in a recently reported paper [41] similarly showed that, when Ir NPs interacted with CeO_2 NPs, Ir-O bonds formed at the interfaces, allowing for electron transfer from Ir atoms to the O_L of CeO_2. The charge density of Ir atoms was adjusted by forming Ir-O bonds between Ir and CeO_2 in order to optimize the adsorption/desorption of intermediates and finally improve the catalytic activity of OER. This phenomenon could be recognized as the primary factor responsible for the observed enhancement in the intrinsic catalytic activity of CeO_2-Ir/CNTs. The durability of OER catalysts is a crucial factor that must be carefully assessed to determine their practicality. To further investigate the stability of CeO_2-Ir/CNTs under the acidic OER conditions, we conducted chronopotentiometry at 10.0 mA cm^{-2}. Figure 3e illustrates that, even after prolonged testing exceeding more than 60.0 h, CeO_2-Ir/CNTs exhibited excellent durability in the acidic OER test with no significant degradation of their catalytic activity. In contrast, c-IrO_2 and Ir/CNTs were completely deactivated within less than 21.0 h and 31.0 h, respectively.

To confirm the high durability of CeO_2-Ir/CNTs in acidic OER tests, the CeO_2-Ir/CNTs' sample after the stability test was characterized using TEM, HRTEM, and XPS. Firstly, TEM analysis confirmed that the morphology of CeO_2-Ir/CNTs remained intact with exceptional structural robustness (Figure 4a). The distribution of CeO_2 and Ir NPs on CNTs could remain uniform (Figure 4b). The paired nanoparticles in Figure 4b retained good crystallinity with lattice fringe spacings of 0.221 nm and 0.192 nm, corresponding to Ir

(111) and CeO_2 (220), respectively (Figure 4c,d). Additionally, XPS characterization was performed on CeO_2-Ir/CNTs to investigate if there was compositional decay during the stability test. Figure 4e–g show that the CeO_2-Ir/CNTs exhibited the XPS spectra of Ir 4f, Ce 3d, and O1s, with the profiles comparable to those before the OER; however, the weaker Ce 3d profile (Figure 4f) may have been attributed to the dissolution of CeO_2. The component analysis revealed a slight decrease in the Ir content of CeO_2-Ir/CNTs after stability testing, while the loss of CeO_2 was more pronounced (Table S4). Therefore, it can be concluded that excellent structural integrity along with stable component composition contributes significantly to maintaining superior activity during acidic OER.

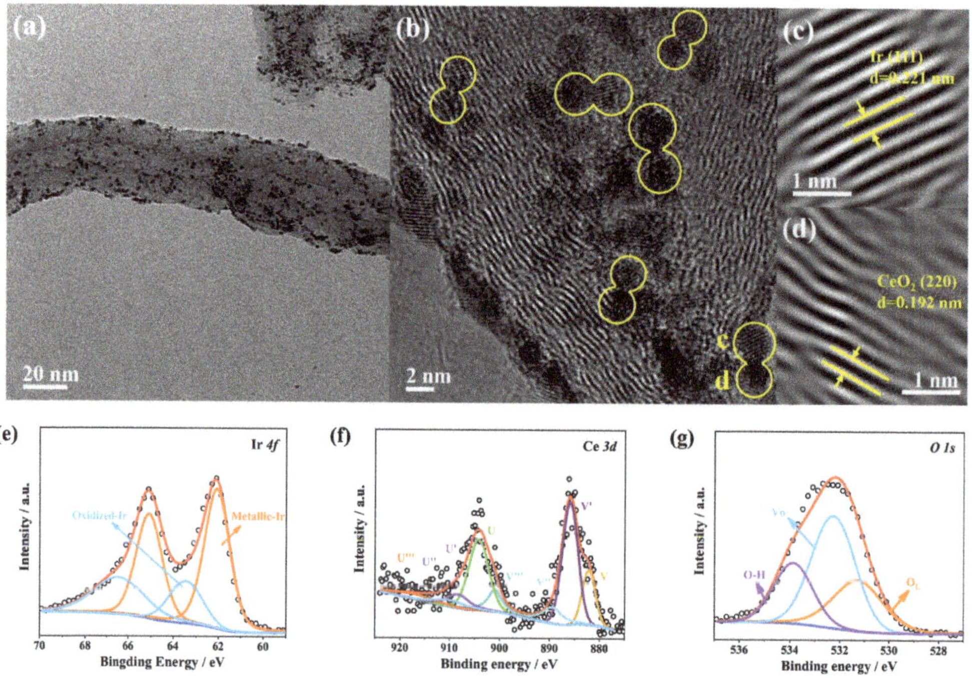

Figure 4. (**a**) TEM, (**b**) HRTEM images, and (**e**–**g**) XPS spectra of CeO_2-Ir/CNTs after OER stability test. (**c**,**d**) represent Ir and CeO_2 nanoparticle lattice fringes from (**b**), respectively.

To further validate the potential application of CeO_2-Ir/CNTs in acidic water electrolyzers, it was evaluated as an anode catalyst for over-all water splitting (OWS). Figure 5a demonstrates the assembly and testing of Pt/C I I CeO_2-Ir/CNTs as the electrode pair in the acidic water electrolyzers. LSVs and chronopotentiometric curves were employed to estimate their OWS performance. Remarkably, a current density of 10.0 mA cm^{-2} could be achieved with an input voltage of only 1.54 V for Pt/C I I CeO_2-Ir/CNTs, while the commercial electrode pair of Pt/C I I c-IrO_2 required a larger input voltage of 1.58 V to achieve the same current density (Figure 5b). The other synthesized samples were assembled with Pt/C to construct electrolyzers and evaluate their performance towards OWS, as depicted in Figure S9. Furthermore, Figure 5c illustrates that Pt/C I I CeO_2-Ir/CNTs could maintain continuous electrolysis for over 60.0 h without significant loss of activity at 10.0 mA cm^{-2}; however, Pt/C I I c-IrO_2 became completely inactive after less than 20.0 h under similar conditions. These results strongly demonstrate the promising potential of CeO_2-Ir/CNTs as efficient anode catalysts for acidic water electrolyzers.

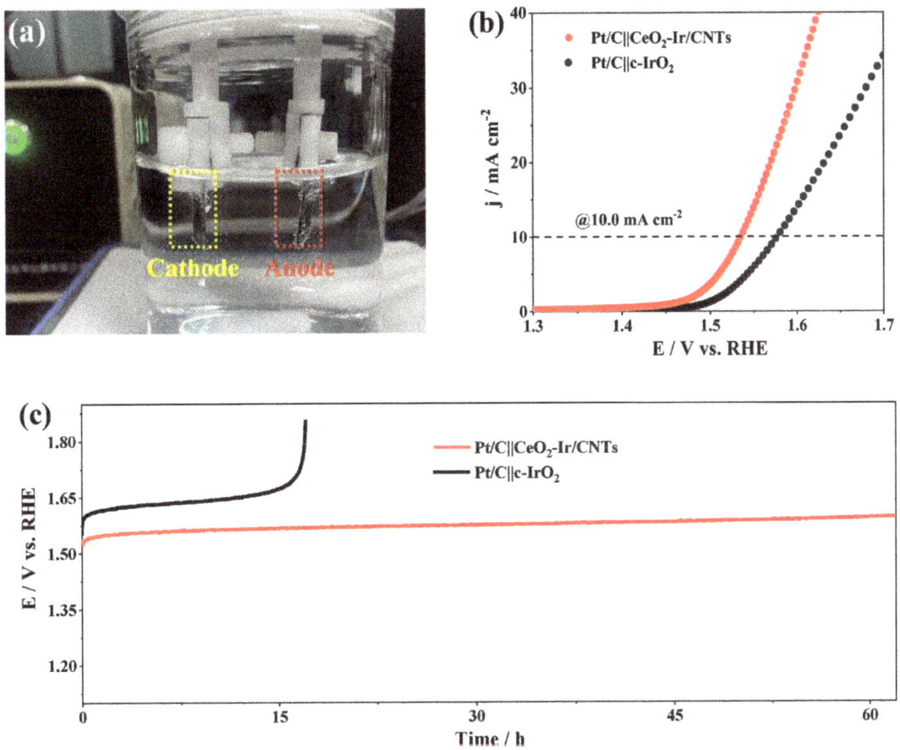

Figure 5. (**a**) Illustration of acidic water electrolyzers. (**b**) LSVs and (**c**) chronopotentiometric curves of Pt/C | | CeO_2-Ir/CNTs and Pt/C | | c-IrO_2 in water electrolyzers using 0.5 M H_2SO_4 as electrolyte.

4. Conclusions

We successfully demonstrated the synthesis of a series of xCeO_2-Ir/CNTs with varying CeO_2 contents using a facile solvothermal method. The catalytic activity of xCeO_2-Ir/CNTs was significantly influenced by the content of CeO_2, with higher catalytic activity observed for acidic OER when the molar ratio of CeO_2/Ir was 1:1. Due to the strong electron coupling in CeO_2-Ir heterojunctions, CeO_2-Ir/CNTs exhibited outstanding OER catalytic performance in acidic media, significantly exceeding the benchmark c-IrO_2. CNTs played a crucial role by providing abundant nucleation sites and excellent conductive pathways for CeO_2-Ir, contributing to the observed high catalytic and mass activity in CeO_2-Ir/CNTs. The acidic water electrolyzers constructed using CeO_2-Ir/CNTs in combination with Pt/C (Pt/C | | CeO_2-Ir/CNTs) demonstrated significant potential for practical applications and outperformed the benchmark c-IrO_2 | | Pt/C system. This study presents a novel approach for the design of efficient, stable, and low-Ir-usage catalysts towards acidic OER.

Supplementary Materials: The following supporting information can be downloaded at: https://www.mdpi.com/article/10.3390/ma16217000/s1, Figure S1: The pH of 0.5 M H_2SO_4 at room temperature; Figure S2: Photograph of catalyst-loaded carbon paper; Figure S3: SEM images of (a) CeO_2-Ir/CNTs, (b) Ir/CNTs and (c) CeO_2/CNTs; Figure S4: XRD pattern of CeO_2/CNTs; Figure S5: (a) XRD patterns, (b) LSVs (solid dot line: with iR compensation), (c) Tafel plots, and (d) EIS Nyquist plots of xCeO_2-Ir/CNTs with different CeO_2 content; Figure S6: (a) The time-dependent volume of O_2 towards OER using CeO_2-Ir/CNTs as anode. The red line represents a fitted line, while square data points denote the volume of O_2 generated at 30-min intervals; Figure S7: CVs recorded at different scan rates in the region of 1.1–1.2 V of (a–d) xCeO_2-Ir/CNTs with different CeO_2 content and (e) c-IrO_2.

(f,g) Cdl plots and ECSAs of the catalysts; Figure S8: ECSA-normalized LSVs of xCeO_2-Ir/CNTs and c-IrO_2; Figure S9: LSVs of Pt/C | | xCeO_2-Ir/CNTs and Pt/C | | CeO_2/CNTs; Table S1: Composition analysis of xCeO_2-Ir/CNTs; Table S2: Analysis of EDS spectrum; Table S3: OER performance comparison of the CeO_2-Ir/CNTs with the catalysts reported in acid media; Table S4: Component analysis of CeO_2-Ir/CNTs before and after OER. References [15,41–53] are cited in the supplementary materials.

Author Contributions: Conceptualization, J.K., M.L. and Y.H.; methodology, J.K., W.L., Z.L. and C.C.; writing—original draft preparation, J.K., Z.L., Y.H. and C.C.; writing—review and editing, J.K., Z.L. and Y.H.; supervision, Y.H. and M.L. All authors have read and agreed to the published version of the manuscript.

Funding: This work was supported by the High-Level Talent Start-Up Fund of Pingdingshan University under Grant PXY-BSQD-202108, the Key Research Projects of Higher Education Institutions in Henan Province under Grant 23A430033, the Young Key Teacher Foundation of Henan Province (2020GGJS227), and the Science and Technology Development Program of Henan province (No. 232102321043).

Institutional Review Board Statement: Not applicable.

Informed Consent Statement: Not applicable.

Data Availability Statement: The authors are able to provide the data presented in this study upon request.

Conflicts of Interest: The authors declare no conflict of interest. The funders had no role in the design of the study, in the collection, analyses, or interpretation of data, in the writing of the manuscript, or in the decision to publish the results.

References

1. Cao, X.; Tan, C.; Zhang, X.; Zhao, W.; Zhang, H. Solution-processed two-dimensional metal dichalcogenide-based nanomaterials for energy storage and conversion. *Adv. Mater.* **2016**, *28*, 6167–6196. [CrossRef] [PubMed]
2. Chen, H.; Cong, T.N.; Yang, W.; Tan, C.; Li, Y.; Ding, Y. Progress in electrical energy storage system: A critical review. *Prog. Nat. Sci.* **2009**, *19*, 291–312. [CrossRef]
3. Chen, Y.; Liu, C.; Xu, J.; Xia, C.; Wang, P.; Xia, B.Y.; Yan, Y.; Wang, X. Key components and design strategy for a proton exchange membrane water electrolyzer. *Small Struct.* **2023**, *4*, 2200130. [CrossRef]
4. Stiber, S.; Balzer, H.; Wierhake, A.; Wirkert, F.J.; Roth, J.; Rost, U.; Brodmann, M.; Lee, J.K.; Bazylak, A.; Waiblinger, W. Porous transport layers for proton exchange membrane electrolysis under extreme conditions of current density, temperature, and pressure. *Adv. Energy Mater.* **2021**, *11*, 2100630. [CrossRef]
5. Li, S.; Gao, Y.; Li, N.; Ge, L.; Bu, X.; Feng, P. Transition metal-based bimetallic MOFs and MOF-derived catalysts for electrochemical oxygen evolution reaction. *Energy Environ. Sci.* **2021**, *14*, 1897–1927. [CrossRef]
6. Li, H.; Lai, J.; Li, Z.; Wang, L. Multi-sites electrocatalysis in high-entropy alloys. *Adv. Funct. Mater.* **2021**, *31*, 2106715. [CrossRef]
7. Wang, X.X.; Swihart, M.T.; Wu, G. Achievements, challenges and perspectives on cathode catalysts in proton exchange membrane fuel cells for transportation. *Nat. Catal.* **2019**, *2*, 578–589. [CrossRef]
8. Rabis, A.; Rodriguez, P.; Schmidt, T. Electrocatalysis for polymer electrolyte fuel cells: Recent achievements and future challenges. *ACS Catal.* **2012**, *2*, 864–890. [CrossRef]
9. Wu, H.; Wang, Y.; Shi, Z.; Wang, X.; Yang, J.; Xiao, M.; Ge, J.; Xing, W.; Liu, C. Recent developments of iridium-based catalysts for the oxygen evolution reaction in acidic water electrolysis. *J. Mater. Chem. A* **2022**, *10*, 13170–13189. [CrossRef]
10. Zhan, Y.; Lu, M.; Yang, S.; Liu, Z.; Lee, J.Y. The origin of catalytic activity of nickel phosphate for oxygen evolution in alkaline solution and its further enhancement by iron substitution. *ChemElectroChem* **2016**, *3*, 615–621. [CrossRef]
11. Luo, F.; Guo, L.; Xie, Y.; Xu, J.; Qu, K.; Yang, Z. Iridium nanorods as a robust and stable bifunctional electrocatalyst for pH-universal water splitting. *Appl. Catal. B* **2020**, *279*, 119394. [CrossRef]
12. Wu, G.; Zheng, X.; Cui, P.; Jiang, H.; Wang, X.; Qu, Y.; Chen, W.; Lin, Y.; Li, H.; Han, X.; et al. A general synthesis approach for amorphous noble metal nanosheets. *Nat. Commun.* **2019**, *10*, 4855. [CrossRef] [PubMed]
13. Zhu, J.-Y.; Xue, Q.; Xue, Y.-Y.; Ding, Y.; Li, F.-M.; Jin, P.; Chen, P.; Chen, Y. Iridium nanotubes as bifunctional electrocatalysts for oxygen evolution and nitrate reduction reactions. *ACS Appl. Mater. Interfaces* **2020**, *12*, 14064–14070. [CrossRef] [PubMed]
14. Chen, X.; Peng, M.; Xiao, D.; Liu, H.; Ma, D. Fully exposed metal clusters: Fabrication and application in alkane dehydrogenation. *ACS Catal.* **2022**, *12*, 12720–12743. [CrossRef]
15. Shan, J.; Guo, C.; Zhu, Y.; Chen, S.; Song, L.; Jaroniec, M.; Zheng, Y.; Qiao, S.-Z. Charge-redistribution-enhanced nanocrystalline Ru@IrO_x electrocatalysts for oxygen evolution in acidic media. *Chem* **2019**, *5*, 445–459. [CrossRef]

16. Liu, X.; Xi, S.; Kim, H.; Kumar, A.; Lee, J.; Wang, J.; Tran, N.Q.; Yang, T.; Shao, X.; Liang, M.; et al. Restructuring highly electron-deficient metal-metal oxides for boosting stability in acidic oxygen evolution reaction. *Nat. Commun.* **2021**, *12*, 5676. [CrossRef] [PubMed]
17. Yang, J.; Shen, Y.; Sun, Y.; Xian, J.; Long, Y.; Li, G. Ir nanoparticles anchored on metal-organic frameworks for efficient overall water splitting under pH-universal conditions. *Angew. Chem. Int. Ed.* **2023**, *135*, e202302220. [CrossRef]
18. Li, Y.; Sun, Y.; Qin, Y.; Zhang, W.; Wang, L.; Luo, M.; Yang, H.; Guo, S. Recent advances on water-splitting electrocatalysis mediated by noble-metal-based nanostructured materials. *Adv. Energy Mater.* **2020**, *10*, 1903120. [CrossRef]
19. Zhao, G.; Jiang, Y.; Dou, S.-X.; Sun, W.; Pan, H. Interface engineering of heterostructured electrocatalysts towards efficient alkaline hydrogen electrocatalysis. *Sci. Bull.* **2021**, *66*, 85–96. [CrossRef]
20. Shi, Z.; Li, J.; Jiang, J.; Wang, Y.; Wang, X.; Li, Y.; Yang, L.; Chu, Y.; Bai, J.; Yang, J.; et al. Enhanced acidic water oxidation by dynamic migration of oxygen species at the Ir/Nb_2O_{5-x} catalyst/support interfaces. *Angew. Chem. Int. Ed.* **2022**, *61*, e202212341. [CrossRef]
21. Wang, J.; Xiao, X.; Liu, Y.; Pan, K.; Pang, H.; Wei, S. The application of CeO_2-based materials in electrocatalysis. *J. Mater. Chem. A* **2019**, *7*, 17675–17702. [CrossRef]
22. Haber, J.A.; Anzenburg, E.; Yano, J.; Kisielowski, C.; Gregoire, J.M. Multiphase nanostructure of a quinary metal oxide electrocatalyst reveals a new direction for OER electrocatalyst design. *Adv. Energy Mater.* **2015**, *5*, 1402307. [CrossRef]
23. Li, C.; Sun, Y.; Djerdj, I.; Voepel, P.; Sack, C.-C.; Weller, T.; Ellinghaus, R.; Sann, J.; Guo, Y.; Smarsly, B.M.; et al. Shape-controlled CeO_2 nanoparticles: Stability and activity in the catalyzed HCl oxidation reaction. *ACS Catal.* **2017**, *7*, 6453–6463. [CrossRef]
24. Wang, Y.; Hao, S.; Liu, X.; Wang, Q.; Su, Z.; Lei, L.; Zhang, X. Ce-doped IrO_2 electrocatalysts with enhanced performance for water oxidation in acidic media. *ACS Appl. Mater. Interfaces* **2020**, *12*, 37006–37012. [CrossRef] [PubMed]
25. Shrestha, N.K.; Patil, S.A.; Seok, J.H.; Salunke, A.S.; Cho, S.; Inamdar, A.I.; Park, Y.; Lee, S.U.; Kim, H.; Im, H. Cerium guided site-selective crystal disorder engineering of MIL-88B (Ni) frameworks for electrocatalysis offering high-performance water oxidation. *Mater. Today Phys.* **2023**, *38*, 101252. [CrossRef]
26. Julkapli, N.M.; Bagheri, S. Graphene supported heterogeneous catalysts: An overview. *Int. J. Hydrog. Energy* **2015**, *40*, 948–979. [CrossRef]
27. Huang, H.; Wang, X. Recent progress on carbon-based support materials for electrocatalysts of direct methanol fuel cells. *J. Mater. Chem. A* **2014**, *2*, 6266–6291. [CrossRef]
28. Ren, J.; Huang, Y.; Zhu, H.; Zhang, B.; Zhu, H.; Shen, S.; Tan, G.; Wu, F.; He, H.; Lan, S.; et al. Recent progress on MOF-derived carbon materials for energy storage. *Carbon Energy* **2020**, *2*, 176–202. [CrossRef]
29. Yan, H.; Jiang, Z.; Deng, B.; Wang, Y.; Jiang, Z.-J. Ultrathin carbon coating and defect engineering promote RuO_2 as an efficient catalyst for acidic oxygen evolution reaction with super-high durability. *Adv. Energy Mater.* **2023**, *13*, 2300152. [CrossRef]
30. Chen, B.; Jiang, Z.; Huang, J.; Deng, B.; Zhou, L.; Jiang, Z.-J.; Liu, M. Cation exchange synthesis of $Ni_xCo_{3-x}O_4$ (x = 1.25) nanoparticles on aminated carbon nanotubes with high catalytic bifunctionality for the oxygen reduction/evolution reaction toward efficient Zn-air batteries. *J. Mater. Chem. A* **2018**, *6*, 9517–9527. [CrossRef]
31. Keller, P.; Strehblow, H.-H. XPS investigations of electrochemically formed passive layers on Fe/Cr-alloys in 0.5 M H_2SO_4. *Corros Sci.* **2004**, *46*, 1939–1952. [CrossRef]
32. Asnavandi, M.; Suryanto BH, R.; Yang, W.; Bo, X.; Zhao, C. Dynamic hydrogen bubble templated NiCu phosphide electrodes for pH-insensitive hydrogen evolution reactions. *ACS Sustain. Chem. Eng.* **2018**, *6*, 2866–2871. [CrossRef]
33. Zhang, B.; Zheng, X.; Voznyy, O.; Comin, R.; Bajdich, M.; García-Melchor, M.; Han, L.; Xu, J.; Liu, M.; Zheng, L. Homogeneously dispersed multimetal oxygen-evolving catalysts. *Science* **2016**, *352*, 333–337. [CrossRef] [PubMed]
34. Xie, X.; Shang, L.; Shi, R.; Waterhouse, G.I.N.; Zhao, J.; Zhang, T. Tubular assemblies of N-doped carbon nanotubes loaded with NiFe alloy nanoparticles as efficient bifunctional catalysts for rechargeable zinc-air batteries. *Nanoscale* **2020**, *12*, 13129–13136. [CrossRef] [PubMed]
35. Pan, X.; Zheng, Z.; Zhang, X.; He, X.; An, Y.; Hao, Y.; Huang, K.; Lei, M. Multi-metallic nanosheets for high-performance hydrogen evolution reaction. *Eng. Sci.* **2022**, *19*, 253–261. [CrossRef]
36. McCrory CC, L.; Jung, S.; Peters, J.C.; Jaramillo, T.F. Benchmarking heterogeneous electrocatalysts for the oxygen evolution reaction. *J. Am. Chem. Soc.* **2013**, *135*, 16977–16987. [CrossRef] [PubMed]
37. Liu, D.; Lv, Q.; Lu, S.; Fang, J.; Zhang, Y.; Wang, X.; Xue, Y.; Zhu, W.; Zhuang, Z. IrCuNi deeply concave nanocubes as highly active oxygen evolution reaction electrocatalyst in acid electrolyte. *Nano Lett.* **2021**, *21*, 2809–2816. [CrossRef]
38. Chen, Y.; Liu, K. Fabrication of Ce/N co-doped TiO_2/diatomite granule catalyst and its improved visible-light-driven photoactivity. *J. Hazard. Mater.* **2017**, *324*, 139–150. [CrossRef]
39. Dhandapani, H.N.; Mahendiran, D.; Karmakar, A.; Devi, P.; Nagappan, S.; Madhu, R.; Bera, K.; Murugan, P.; Babu, B.R.; Kundu, S. Boosting of overall water splitting activity by regulating the electron distribution over the active sites of Ce doped NiCo–LDH and atomic level understanding of the catalyst by DFT study. *J. Mater. Chem. A* **2022**, *10*, 17488–17500. [CrossRef]
40. Möller, S.; Barwe, S.; Masa, J.; Wintrich, D.; Seisel, S.; Baltruschat, H.; Schuhmann, W. Online monitoring of electrochemical carbon corrosion in alkaline electrolytes by differential electrochemical mass spectrometry. *Angew. Chem. Int. Ed.* **2020**, *59*, 1585–1589. [CrossRef]

41. Chen, X.; Liao, W.; Zhong, M.; Chen, J.; Yan, S.; Li, W.; Wang, C.; Chen, W.; Lu, X. Rational design of robust iridium-ceria oxide-carbon nanofibers to boost oxygen evolution reaction in both alkaline and acidic media. *Nano Res.* **2023**, *16*, 7724–7732. [CrossRef]
42. Gao, J.; Xu, C.-Q.; Hung, S.-F.; Liu, W.; Cai, W.; Zeng, Z.; Jia, C.; Chen, H.M.; Xiao, H.; Li, J.; et al. Breaking long-range order in iridium oxide by alkali ion for efficient water oxidation. *J. Am. Chem. Soc.* **2019**, *141*, 3014–3023. [CrossRef] [PubMed]
43. Li, G.; Li, K.; Yang, L.; Chang, J.; Ma, R.; Wu, Z.; Ge, J.; Liu, C.; Xing, W. Boosted performance of Ir species by employing TiN as the support toward oxygen evolution reaction. *ACS Appl. Mater. Interfaces* **2018**, *10*, 38117–38124. [CrossRef] [PubMed]
44. Wu, X.; Feng, B.; Li, W.; Niu, Y.; Yu, Y.; Lu, S.; Zhong, C.; Liu, P.; Tian, Z.; Chen, L.; et al. Metal-support interaction boosted electrocatalysis of ultrasmall iridium nanoparticles supported on nitrogen doped graphene for highly efficient water electrolysis in acidic and alkaline media. *Nano Energy* **2019**, *62*, 117–126. [CrossRef]
45. Zhang, H.; Shi, P.; Ma, X.; Ma, C.; Han, S.; He, C.; Wu, H.; Zhu, L.; Zhang, B.; Lu, Y.; et al. Construction of ordered atomic donor-acceptor architectures in bcc IrGa intermetallic compounds toward highly electroactive and stable overall water splitting. *Adv. Energy Mater.* **2022**, *13*, 2202703. [CrossRef]
46. Zhang, L.; Jang, H.; Li, Z.; Liu, H.; Kim, M.G.; Liu, X.; Cho, J. $SrIrO_3$ modified with laminar Sr_2IrO_4 as a robust bifunctional electrocatalyst for overall water splitting in acidic media. *Chem. Eng. J.* **2021**, *419*, 129604. [CrossRef]
47. Zheng, X.; Qin, M.; Ma, S.; Chen, Y.; Ning, H.; Yang, R.; Mao, S.; Wang, Y. Strong oxide-support interaction over IrO_2/V_2O_5 for efficient pH-universal water splitting. *Adv. Sci.* **2022**, *9*, 2104636. [CrossRef]
48. Zhang, Y.; Zhang, G.; Zhang, M.; Zhu, X.; Shi, P.; Wang, S.; Wang, A.-L. Synergistic electronic and morphological modulation by trace Ir introduction boosting oxygen evolution performance over a wide pH range. *Chem. Eng. J.* **2022**, *433*, 133577. [CrossRef]
49. Seitz, L.C.; Dickens, C.F.; Nishio, K.; Hikita, Y.; Montoya, J.; Doyle, A.; Kirk, C.; Vojvodic, A.; Hwang, H.Y.; Norskov, J.K.; et al. A highly active and stable $IrO_x/SrIrO_3$ catalyst for the oxygen evolution reaction. *Science* **2016**, *353*, 1011–1014. [CrossRef]
50. Chen, X.; Li, W.; Song, N.; Zhong, M.; Yan, S.; Xu, J.; Zhu, W.; Wang, C.; Lu, X. Electronic modulation of iridium-molybdenum oxides with a low crystallinity for high-efficiency acidic oxygen evolution reaction. *Chem. Eng. J.* **2022**, *440*, 135851. [CrossRef]
51. Nong, H.N.; Reier, T.; Oh, H.-S.; Gliech, M.; Paciok, P.; Vu, T.H.T.; Teschner, D.; Heggen, M.; Petkov, V.; Schlögl, R.; et al. A unique oxygen ligand environment facilitates water oxidation in hole-doped $IrNiO_x$ core–shell electrocatalysts. *Nat. Catal.* **2018**, *1*, 841–851. [CrossRef]
52. Zhu, M.; Shao, Q.; Qian, Y.; Huang, X. Superior overall water splitting electrocatalysis in acidic conditions enabled by bimetallic Ir-Ag nanotubes. *Nano Energy* **2019**, *56*, 330–337. [CrossRef]
53. Grimaud, A.; Demortiere, A.; Saubanere, M.; Dachraoui, W.; Duchamp, M.; Doublet, M.-L.; Tarascon, J.-M. Activation of surface oxygen sites on an iridium-based model catalyst for the oxygen evolution reaction. *Nat. Energy* **2016**, *2*, 16189. [CrossRef]

materials

Article

Synthesis and Catalytic Performance of Mo_2C/MoS_2 Composite Heterojunction Catalysts

Congyi Zhang [1], Zhigang Pan [1,2] and Yaqiu Tao [1,2,*]

1 College of Materials Science and Engineering, Nanjing Tech University, Nanjing 211800, China; 202161203161@njtech.edu.cn (C.Z.); panzhigang@njtech.edu.cn (Z.P.)
2 State Key Laboratory of Materials-Oriented Chemical Engineering, Nanjing 211800, China
* Correspondence: taoyaqiu@njtech.edu.cn; Tel.: +86-137-7078-0496

Abstract: Hydrogen, as a clean, safe, and efficient energy carrier, is one of the hot energy sources that have attracted much attention. Mo_2C, due to the introduction of C atoms, makes the atomic spacing of the Mo lattice decrease and changes the width of the d-band, which makes the electronic properties of Mo_2C similar to that of Pt noble metals, exhibiting excellent electrochemical hydrogen precipitation performance. MoS_2, due to its special crystal structure and tunable electronic structure, has been widely studied. In this paper, Mo_2C nanoparticles were prepared by high-temperature carbonization, and then two-dimensional layered MoS_2 were be loaded on Mo_2C nanoparticles by the hydrothermal method to synthesize Mo_2C/MoS_2 composite catalysts. Their electrochemical hydrogen precipitation (HER) performance under acidic conditions was tested. The above catalysts were also characterized by modern material testing methods such as XRD, SEM, TEM, and XPS. The results showed that the composite catalysts exhibited the most excellent electrochemical hydrogen precipitation performance at Mo_2C/MoS_2-3, with the lowest overpotential at a current density of 10 mA cm^{-2}, Tafel slope, and electrochemical impedance. At the same time, the electrochemically active area was dramatically enhanced, with good stability under prolonged testing. The catalytic activity was significantly improved compared with that of Mo_2C and MoS_2. The characterization and experimental results indicate that the heterogeneous structure of Mo_2C and MoS_2 formed a built-in electric field between the two, which accelerated the electron transfer efficiency and provided more active sites. The Mo_2C/MoS_2 composite catalyst is a low-cost, easy-to-prepare, and high-efficiency electrochemical hydrogen precipitation catalyst, providing a new idea for developing green and clean energy.

Keywords: Mo_2C; MoS_2; hetrostructure; HER

1. Introduction

With the continuous development of society, humanity's demand for energy continues to expand, and the excessive use of traditional fossil fuels has led to environmental problems and the emergence of the crisis of energy scarcity. Here, there is the search for a new type of clean energy to eliminate the dependence on the limited fossil fuels for the sustainable development of humanity, where social stability and progress of the problems faced have an urgent need to be resolved [1].

Hydrogen has emerged as a paragon of clean energy carriers, embodying a prodigious potential for high energy yield per mass (142 MJ kg^{-1}), dwarfing even the venerable fossil fuels, bolstering its prominence at the pinnacles of energy research. The gamut of prevalent hydrogen fabrication techniques encompasses photocatalytic [2,3] and electrolytic hydrogen generation [4–6], biomass enzymatic degradation [7,8], and conventional fossil-fuel-based hydrogen extraction [9,10]. For example, Marjeta et al. [11] synthesized efficient photocatalysts by epitaxial growth of $SrTiO_3$ on $Bi_4Ti_3O_{12}$ substrates. The multilayer composite catalysts CoS_2/MoS_2 and Bi_2S_2/MoS_2 prepared by Yang et al. [12] exhibit high catalytic activity with high specific surface area and unique surface morphology. Of these modalities, the electrolytic conjuration of hydrogen is lauded for its impeccability and environmental benevolence, juxtaposed with the offscourings of its counterparts.

Electrochemical hydrogen evolution is a process whereby hydrogen gas is produced through the directional charge transfer in an electrochemical reaction. Within the electrode system, cations in the electrolyte migrate towards the cathode, absorbing electrons and undergoing a reduction reaction to generate hydrogen gas; anions in the electrolyte move towards the anode, release electrons, and undergo an oxidation reaction to generate oxygen gas. This method boasts advantages such as efficiency, environmental friendliness, and safety. In electrochemical hydrogen evolution, the electrodes are key in catalyzing the reaction, commonly utilizing transition metal compounds such as platinum, palladium, iron, and nickel, which demonstrate excellent catalytic activity and can significantly reduce the potential required for the electrochemical hydrogen evolution reaction. Molybdenum-based compounds are also among the hot materials currently under research as catalysts for electrochemical hydrogen evolution. The electrolyte solution is typically acidic (0.5 mol L^{-1} H_2SO_4) or alkaline (1 mol L^{-1} KOH solution).

Water electrolysis hydrogen evolution is divided into the hydrogen evolution reaction (HER) and the oxygen evolution reaction (OER) [13], each occurring at the cathode and anode of the electrode, respectively. HER is one of the simplest electrochemical reactions, involving a two-electron transfer; OER, on the other hand, involves a four-electron transfer. Compared to OER, HER occurs more easily and is more efficient, with the required over-potential being lower. Therefore, water electrolysis hydrogen evolution is a simple and efficient hydrogen production strategy. The reaction equation for HER is

$$2H^+ + 2e^- \rightarrow H_2 \tag{1}$$

This reaction consists of three steps: the Volmer, Tafel, and Heyrovsky reactions [14–16], which correspond to different stages in the electrochemical hydrogen evolution process. The Volmer reaction is considered the initial step of the electrochemical hydrogen evolution reaction and is the only step considered in the theoretical model of the single-electron HER. In the Volmer reaction, H^+ ions adsorb onto the catalyst surface to form adsorbed hydrogen atoms (H^*), where * indicates an active site on the catalyst. The Tafel reaction is where two adsorbed H^* atoms combine and desorb to form H_2. The Heyrovsky reaction involves an adsorbed H^* combining an H^+ ion from the electrolyte and an electron e^- to desorb and form H_2. In alkaline conditions, the required H^* is produced through the dissociation of H_2O molecules, and hence the reaction equation varies.

Volmer $$H^+ + e^- \rightarrow H^* \text{ (acid)} \tag{2}$$

Volmer $$H_2O + e^- \rightarrow H^* + OH^- \text{ (alkaline)} \tag{3}$$

Heyrovsky $$H^* + H^+ + e^- \rightarrow H_2 \text{ (acid)} \tag{4}$$

Heyrovsky $$H_2O + H^* + e^- \rightarrow H_2 + OH^- \text{ (alkaline)} \tag{5}$$

Tafel $$H^* + H^* \rightarrow H_2 \tag{6}$$

In HER, the Tafel slope is usually used to determine the mechanism of cathodic H_2 generation: since the HER reaction necessarily includes the process of electron transfer and desorption, when the Tafel slope is around 120 mV dec^{-1}, the decisive step of HER is the Volmer reaction, and the electron transfer rate is slow, so the reaction rate can be accelerated by increasing the active sites of the catalyst at this time. When the Tafel slope is around 40 mV dec^{-1}, the decisive step of HER is the Heyrovsky reaction, the desorption process is slow, and the reaction rate can be accelerated by changing the physical properties of the catalyst (specific surface area, roughness, etc.). When the Tafel slope is around 30 mV dec^{-1}, the electron transfer and desorption rate on the electrode surface are very fast, and then the reaction decisive step is the compound desorption process of H*, i.e., the Tafel reaction, which can be accelerated by increasing the active area of the catalyst as well.

Platinum on carbon (Pt/C) has historically reigned as the superlative catalyst for electrochemical hydrogen production [17], acclaimed for its exceptional efficacy. However, the prohibitive cost of platinum, coupled with its scant availability and vulnerability to catalyst poisoning, invites the imperative of conceiving economical, facilely synthesized, robust, safe, and innocuous catalytic surrogates. Amidst this pursuit, an array of transition metal derivatives, including oxides, carbides, and sulfides, have emerged as contenders exhibiting laudable electrochemical propensities. Liu et al. [18] doped Ru single atoms into CoP nanoparticles and then loaded this catalyst on carbon dots (CDs) to form Ru-CoP/CDs composites. The CDs, which were made from citric acid and ethylenediamine, could regulate the electrocatalytic reaction sites and electronic structure of Ru-CoP/CDs by inhibiting the growth and agglomeration of Ru-CoP, resulting in excellent electrocatalytic performance of the catalysts. In 2021, Park et al. [19] improved the catalytic activity by introducing Ni to form a Ni-Mo binary alloy nitride (Ni_2Mo_3N) to weaken the Mo-H bonding and thus improve the catalytic activity, which enabled the catalyst to exhibit excellent catalytic performance. Notably, molybdenum carbide (Mo_2C) has garnered significant interest as a material of meritorious distinction among electrocatalysts.

Molybdenum carbide [20], a venerated constituent within the pantheon of transition metal carbides, manifests an engorged metallic framework and an augmented interatomic chasm among Mo atoms. This structural transformation mitigates orbital congruency among neighboring metal atoms. Insights from Density Functional Theory (DFT) intimations denote that the electron conjugation across the d orbitals of Mo and the s and p orbitals of carbon engenders a marked augmentation of the d-band in metallic Mo [21]. Such alterations consummate an electronic schema and catalytic comportment mirroring that of the noble metals partaking of Group VIII, thus ensuring a pronounced HER acumen [22].

Utilizing the two-dimensional layered material molybdenum disulfide (MoS_2), notable for its adjustable electronic properties, robust catalytic stability, and significant natural abundance, MoS_2 emerges as an exceptional electrocatalyst [23].

Extant literature underscores that monolayer molybdenum disulfide, a consummate archetype of transition metal dichalcogenides (TMDCs), has garnered extensive scrutiny owing to its substantial in-plane electron mobility, pronounced current switching ratio, superior photoelectrocatalytic attributes, and robust catalytic hydrogen evolution activity. Of the trio of molybdenum disulfide allotropes (1T, 2H, and 3R), the 2H phase prevails, owing to its augmented stability, crystallizing into a hexagonal lattice with a pronounced planar stratification. Within the MoS_2 architecture, Mo atoms are conjoined with S atoms via ionic linkages, whilst S–S interactions are governed by Van der Waals forces, producing a unique sandwich-like configuration that finds application across industries such as aerospace, lubrication, and energy storage. The 1T-MoS_2, intrinsically metallic, adopts a tetragonal crystalline assembly, and while it is catalytically active, it is inherently unstable, typically transforming to the more stable 2H phase upon annealing. Conversely, the 3R phase of MoS_2, which crystallizes under high temperatures within the rhombohedral system, is prone to a phase transition into the 2H-MoS_2 configuration. The preponderance of contemporary research in electrochemical hydrogen evolution orbits around exploring the 2H phase [24,25].

In pursuit of catalytic interfaces conducive to electrochemical hydrogen production—emphasizing potent catalytic efficacy, enduring stability, and economic viability—this investigation involves the synthesis of molybdenum carbide (Mo_2C) nanoparticles via a high-temperature carbonization route, segueing into the generation of Mo_2C/MoS_2 composite heterojunction semiconductor catalysts through the integration of two-dimensional MoS_2 onto Mo_2C nanoparticles using hydrothermal methodology. Employing advanced material characterization techniques such as X-ray diffraction (XRD), scanning electron microscopy (SEM), and X-ray photoelectron spectroscopy (XPS), these catalysts were meticulously examined. Their electrochemical hydrogen evolution capability was subsequently evaluated by employing a CHI660E electrochemical workstation. The synthesis revealed that amalgamating Mo_2C with MoS_2 considerably ameliorates the electronic structure at both the surface and interface of the catalyst. In the composite, n-type Mo_2C operates as an electron reservoir, augmenting the electrical conductivity within the layered structure and concurrently unveiling additional active sites. Notably, the hydrogen adsorption-free energy (ΔG_H) of Mo_2C is discernibly less negative (−0.65 eV) compared to MoS_2 (2.05 eV), underscoring the enhanced propensity of Mo_2C to adhere to H^+ and conduct the reduction reaction with heightened efficiency [26,27]. The bifurcation of components also culminates in an optimized electrochemical profile for the catalyst. Electrochemical assays ascertain that the composite catalysts outperform isolated Mo_2C and MoS_2 nanoparticles in terms of hydrogen evolution. Further optimization of the electronic structure was achieved by fine-tuning the molar ratio of the constituent materials, resulting in an elevation of catalytic performance. In this study, a composite catalyst with a core-shell structure was prepared by innovatively growing MoS_2 encapsulated on the surface of Mo_2C grains via a facile synthesis method, which exhibited significantly better electrocatalytic performance than that of Mo_2C and MoS_2, as well as showing good stability over a long-time test.

2. Materials and Method

2.1. Materials

Thiourea (CH_4N_2S) was purchased from Tianjin Zhiyuan Chemical Reagent Co. (Tianjin, China). Ammonium molybdate tetrahydrate ($(NH_4)_6Mo_7O_{24}\cdot 4H_2O$, 99%) and 1,8-diaminonaphthalene ($C_{10}H_{10}N_2$) were purchased from Sinophosphate Chemical Reagent Co. (Tianjin, China). Hydrochloric acid (HCl) and sulfuric acid (H_2SO_4) were supplied by Nanjing Chemical Specimen Co. (Nanjing, China). All chemicals are analytical grade and used without further purification.

2.2. Synthesis of Electrocatalysts

Mo_2C flakes: In this experiment, pure β-Mo_2C was prepared by the calcination method. Ammonium molybdate tetrahydrate ($(NH_4)_6Mo_7O_{24}\cdot 4H_2O$) was used as the molybdenum source, and 1,8-diaminonaphthalene ($C_{10}H_{10}N_2$) was used as the carbon source. The calcination precursor was prepared by mixing 2 mmol of ammonium molybdate tetrahydrate in 12 mL of deionized water and 3.2 mmol of 1,8-diaminonaphthalene in 10 mL of anhydrous ethanol. The pH of the mixture was adjusted by 1 mol L^{-1} hydrochloric acid to obtain the precipitation with a molar of 1:2.25 for n (Mo):n (C). The solution was left for 30 min, and then the precipitation was collected and dried at 100 °C for 10 h to obtain the Mo_2C precursor. The precursor was ground into powder and calcinated in a vacuumed high-temperature tube furnace, which was heated up to 725 °C at a rate of 5 °C min^{-1} and kept warm for 4 h. After that, it was naturally cooled down to room temperature to obtain β-Mo_2C nanoparticles.

MoS_2 particles: Ammonium molybdate tetrahydrate was used as the molybdenum source and thiourea (CH_4N_2S) as the sulfur source. Typically, 2 mmol ammonium molybdate tetrahydrate and 5.5 mmol thiourea were dissolved in 60 mL of deionized water with stirring to keep a molar ratio of n(Mo): n(S) = 1:4. The black suspension was then transferred into the polytetrafluoroethylene liner of a stainless steel high-temperature hydrothermal reactor. The black suspension was filtered after being heated at 200° for 8 h to obtain the black solid powder, which was washed three times with 95% ethanol and distilled water and then dried in a constant temperature blast oven at 80 °C for 8 h to obtain the MoS_2 nanoparticles.

Mo_2C/MoS_2 composite catalyst: Mo_2C nanoparticles were prepared using the above method. The composite catalysts with different Mo_2C/MoS_2 molar ratios (0.5, 1, 2, 3, and 4) were prepared by the same hydrothermal method using ammonium molybdate tetrahydrate as the molybdenum source and thiourea as the sulfur source, keeping the content of Mo_2C unchanged.

2.3. Characterization

Powder X-ray diffraction patterns were obtained by a Rigaku SmartLab diffractometer (Tokyo, Japan) using Cu Kα (λ = 0.154178 nm) as a radiation source. The morphology of the samples was measured by scanning electron microscope (FESEM, Ultra-55, Zeiss, Oberkochen, Germany). The samples were pretreated at 105 °C for 12 h before measurement. The electrochemical impedance was measured at the electrochemical workstation CH1660E. X-ray photoelectron spectroscopy was recorded using Kratos AXIS Supra (Shimadzu, Kyoto, Japan).

2.4. Electrochemical Test

In this experiment, electrochemical tests were performed using a standard three-electrode system to evaluate the electrochemical hydrogen precipitation performance of the catalyst. The CHI660E electrochemical workstation was used to load the powder samples to be tested onto a glassy carbon electrode of diameter ϕ = 3 mm with 5 wt% concentration of Nafion solution at 3.5 mg cm^{-2}. The electrode prepared as described above was used as the working electrode, the C electrode was used as the counter electrode, and Ag/AgCl (saturated KCl-filled) was used as the reference electrode. The LSV curves were obtained by reversible hydrogen electrode potential conversion, and the potential conversion equation for Ag/AgCl was

$$E(vs.\ RHE) = E(vs.\ \mathrm{Ag/AgCl}) + 0.0592pH + 0.197 \quad (7)$$

The electrolyte was a 0.5 mol L^{-1} H_2SO_4 solution and linear voltammetry (LSV) was tested at a sweep rate of 5 mV s^{-1}. The electrochemical active area (ECSA) of the catalysts was evaluated by testing the double layer capacitance (C_{dl}), which was subjected to cyclic voltammetry (CV) at sweep rates of 20 mV s^{-1}, 40 mV s^{-1}, 60 mV s^{-1}, 80 mV s^{-1}, and 100 mV s^{-1} at potentials ranging from 0.15 to 0.25 V. The catalysts were tested at a sweep rate of 5 mV s^{-1}. The electrochemical impedance (EIS) of the catalyst was measured in the frequency range of 100,000–0.1 Hz to evaluate the electrochemical hydrogen precipitation performance of the catalyst. The stability of the catalysts was evaluated by recording the changes in the LSV curves before and after 1000 cycles of CV testing and recording the curves for 12 h. EIS was fitted with the program Zview and all experimental results are reproducible.

3. Results and Discussion

3.1. Characterization of Electrocatalysts

Figure 1 elegantly delineates the X-ray diffraction (XRD) spectra of Mo_2C; MoS_2; and their composite, Mo_2C/MoS_2, catalysts. Within these spectra, distinct peaks at 34.4°, 37.9°, and 39.5° are observable, corresponding, respectively, to the (100), (002), and (101) lattice planes of β-Mo_2C crystals. These peaks are congruent with the salient features of the β-Mo_2C standard PDF card (PDF#35-0708). The MoS_2 sample, synthesized via hydrothermal methods, exhibited characteristic peaks at 13.5° and 32.4°. These are indicative of the (002) and (100) facets of MoS_2 crystals, thus confirming the efficacy of the hydrothermal synthesis in crafting MoS_2. However, it can be observed that the position of the MoS_2 diffraction peak was low compared to the 2θ value corresponding to the standard peak of 2H-phase MoS_2, indicating the existence of multiple crystalline phases in the synthesized MoS_2 samples. 1T-phase MoS_2 requires lower reaction temperatures and shorter reaction times compared to the 2H-phase MoS_2, and the position of its (002) crystalline diffraction peaks was around 9°, which resulted in the leftward diffraction peak position. Nevertheless, a diminution in peak intensity is noted in the XRD pattern of the Mo_2C/MoS_2 composite due to the 2D layered structure and diminutive grain size of MoS_2. This nuanced two-dimensional structure manifested less intense diffraction peaks with broader half-peak breadths in pristine MoS_2 samples. In contrast, β-Mo_2C displayed pronounced, acutely defined peaks due to enlarged grain size and the formation of nanorods through the high-temperature carbonization of molecular clusters. Consequently, in the composite samples, β-Mo_2C's peaks dwarfed those of MoS_2 in intensity. Yet, vestiges of MoS_2's diffraction peaks at 13.5° and 32.4° remained discernible, substantiating the coexistence of dual-phase Mo_2C and MoS_2 within the composite. Importantly, the absence of extraneous stray peaks corroborates the sample's compositional purity.

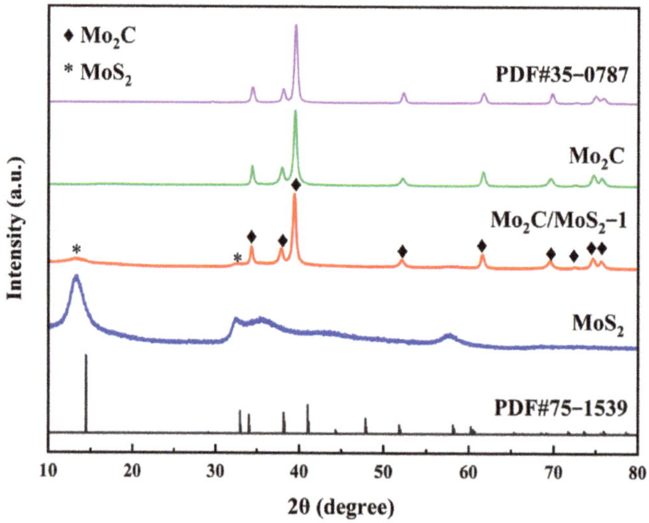

Figure 1. XRD pattern of Mo_2C, MoS_2, and Mo_2C/MoS_2.

Analysis of the X-ray photoelectron spectroscopy (XPS) spectra reveals the distinctive peaks at 231.5 eV and 227.9 eV in Figure 2a, which are attributed to the $3d_{3/2}$ and $3d_{5/2}$ orbitals of divalent molybdenum (Mo^{2+}) in Mo_2C, respectively. Conversely, peaks observed at 235.2 eV and 231.2 eV correspond to the $3d_{3/2}$ and $3d_{5/2}$ orbitals of hexavalent molybdenum (Mo^{6+}) in MoO_3 oxides, signifying oxidative processes affecting the sample surfaces during both the reaction phase and preservation. The Mo_2C/MoS_2-1 composite catalysts manifested characteristic peaks at 230.6 eV and 227.4 eV, indicative of the Mo^{2+} $3d_{3/2}$ and $3d_{5/2}$ orbitals of the Mo element. These peaks were shifted by 0.53 eV to lower binding energies relative to those of Mo_2C alone, suggesting an augmented electron cloud density around the composite samples [28]. This phenomenon correlates with a decrease in electrochemical impedance and an enhancement in electrocatalytic performance. Figure 2b corroborates the findings from electrochemical characterization. Peaks at 230.6 eV and 227.4 eV correspond to the presence of tetravalent molybdenum (Mo^{4+}) $3d_{3/2}$ and $3d_{5/2}$ orbitals characteristic of MoS_2. In contrast, peaks at 236.0 eV and 232.2 eV are ascribed to MoO_3, while those at 233.6 eV and 230.0 eV denote an intermediary oxidation state of molybdenum ($Mo^{\varepsilon+}$) with $4 < \varepsilon < 6$. The intermediate oxidation state, as evidenced by the peaks, indicates that the MoS_2 samples also experienced surface oxidation; notably, the peak at 226.6 eV originated from S2s interference.

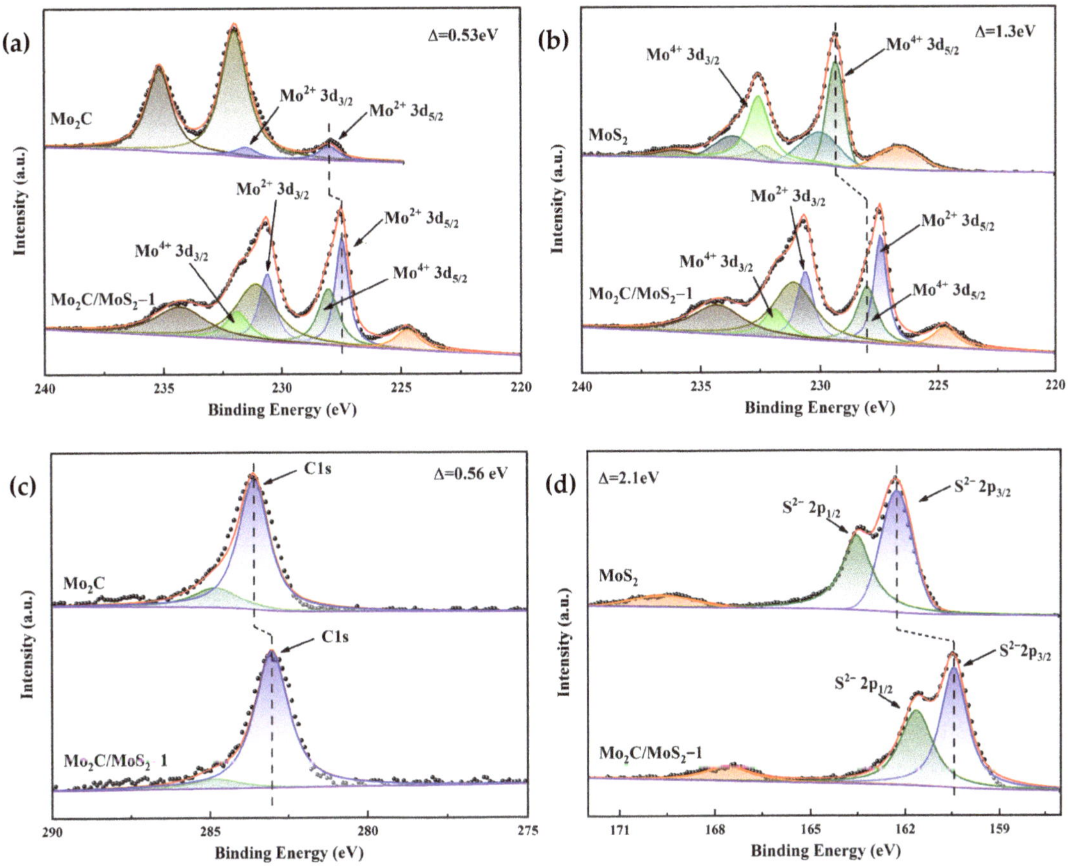

Figure 2. XPS images of Mo_2C, MoS_2, and Mo_2C/MoS_2-1 composite catalysts: (**a**,**b**) Mo3d; (**c**) C1s; (**d**) S2p.

In the Mo_2C/MoS_2-1 samples, the peaks at 231.8 eV and 228.0 eV are ascribed to MoS_2 components' Mo^{4+} $3d_{3/2}$ and $3d_{5/2}$ orbitals. They shifted toward lower binding energy relative to the MoS_2 sample by approximately 1.3 eV. This shift suggests that the electron cloud density around the MoS_2 component increased upon composite formation, with the Mo_2C component transferring more free electrons to the MoS_2 due to Mo_2C's role as an electron donor, enriching the electron milieu surrounding MoS_2. The binding energy discrepancies, denoted by Δ = 3.2 eV for Mo^{2+} spin-split orbitals and Δ = 3.8 eV for Mo^{4+} spin orbitals, differed from those in the Mo_2C (Δ = 3.6 eV) and MoS_2 (Δ = 3.2 eV) samples, respectively. These differences indicate a successful heterostructure of the composites, leading to an optimized electronic architecture conducive to superior catalytic performance [29]. The shifting to lower binding energies is also witnessed in the characteristic C1s peaks at 283.0 eV for Mo_2C/MoS_2 and 283.6 eV for Mo_2C, as noted in Figure 2c. Lastly, Figure 2d showcases peaks at 161.6 eV and 160.4 eV for the Mo_2C/MoS_2-1 samples, corresponding to the $S2^{-2}$ $p_{1/2}$ and $S2^{-2}$ $p_{3/2}$ orbitals, respectively. These peaks were discernibly shifted towards lower binding energies compared to the 163.5 eV and 162.2 eV peaks of MoS_2, embodying the changing electron densities in these subsidized structures.

The specimens underwent meticulous examination utilizing scanning electron microscopy (SEM), with the corresponding findings delineated in Figure 3. Observational analysis of Figure 3a reveals that molybdenum carbide (Mo_2C) assumed an elongate ellipsoidal morphology, with each particle approximately 80 nm in diameter, which coalesced to constitute substantial nanorods. These nanorods exhibited a porous exterior, spanning sizes from 400 to 600 nm. Subsequent inspection of Figure 3b unveils a scanning electron micrographic depiction of molybdenum disulfide (MoS_2), which displays a composition of stratified nanofloral configurations. The distinctly layered petal-like structure corroborates the dimensional stratification intrinsic to MoS_2. Upon amalgamation of the two entities, as depicted in Figure 3c, the Mo_2C phase receded nearly beyond detection; nonetheless, one can discern the presence of nanofloral spherical entities, ranging in size from 600 to 800 nm. This indicates a transformative alteration of MoS_2's inherent cuboidal morphology post-composite, with the spherical nanofloral architecture potentially unveiling an augmented array of reactive sites. Figure 3d provides a vista into the amalgamated growth of Mo_2C and MoS_2 grains through mutual intercalation. Post-combination morphological transformation yielded predominant agglomerations of spherical Mo_2C particles, the surfaces of which were adorned with the nano-floral constructions of MoS_2, substantiating a successful synthesis of Mo_2C/MoS_2 heterojunction semiconductor catalysts.

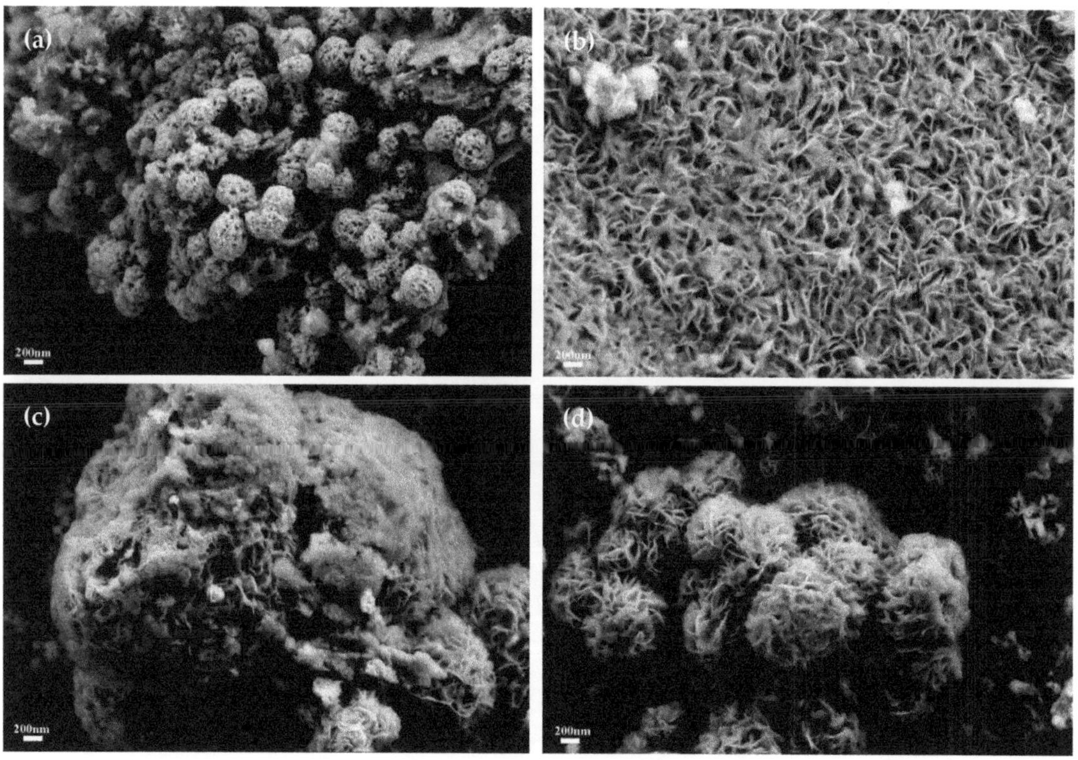

Figure 3. SEM images of catalyst samples: (**a**) Mo_2C; (**b**) MoS_2; (**c**,**d**) Mo_2C/MoS_2.

In the micrographs obtained via transmission electron microscopy of the Mo_2C/MoS_2 heterojunction catalysts, one can discern the characteristic striping of MoS_2 lamellae interspersed with Mo_2C crystallites. The MoS_2 had a predilection for ensheathing the Mo_2C nuclei, engulfing them in a continuous veil that contributed to the predominant MoS_2 crystalline façade evident in scanning electron microscopy depictions. A proliferation of defects within the superficial MoS_2 layers festooning the Mo_2C granules correlated with an augmented exhibition of dynamic sites, consequently enhancing the catalyst's overall reactivity. The high-resolution transmission electron microscopy delineation in Figure 4b revealed lattice fringes with an interplanar distance of 0.62 nm, denoting the (002) planes of MoS_2 crystals, while those fringes with a separation of 0.24 nm are ascribed to the (101) planes of Mo_2C crystals—a measurement marginally exceeding the canonical value of 0.228 nm. This subtle expansion intimates a tensile distortion within the Mo_2C matrix as occasioned by the admixture with MoS_2, a hallmark of the heterostructures' synthesis through the successful amalgamation of the constituent Mo_2C and MoS_2 materials.

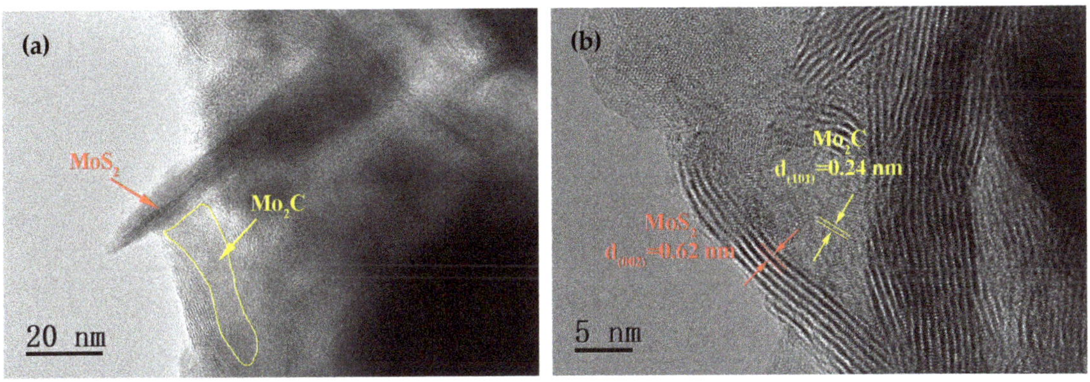

Figure 4. TEM image of Mo_2C/MoS_2 composite catalysts: (**a**) Mo_2C/MoS_2 composite catalyst; (**b**) localized enlarged view.

3.2. Electrochemical Results of Electrocatalysts

Figure 5 depicts the electrochemical evaluation results for the various specimens. As evidenced by Figure 5a,b, the Mo_2C/MoS_2 composite catalyst exhibited minimal overpotential at an identical current density. Specifically, this catalyst manifested an overpotential of merely 278 mV at 10 mA cm^{-2} current density, thereby signifying its superior hydrolytic proficiency for electrochemical hydrogen evolution in contrast to the pristine specimen. Concurrently, the Tafel inclination for the Mo_2C/MoS_2 composite experienced a considerable descent to 136 mV dec^{-1}, markedly lower than the 189 mV dec^{-1} for undiluted Mo_2C and the 323 mV dec^{-1} for unadulterated MoS_2. This observation implies that the composite catalyst substantially diminished the activation energy requisite for the hydrolytic process when paralleled with the pure specimen. Furthermore, Figure 5c unveils the electrochemical impedance spectroscopy (EIS) representations for the unmingled Mo_2C, MoS_2, and the composite catalyst. The equivalent circuit diagram and specific values are shown in Table 1. It becomes apparent that the composite samples exhibited a reduced arc radius within the high-frequency region than their unmingled counterparts, a factor that intimates a lower impedance and an accelerated rate of electron transference. This enhancement facilitates continuous electron conveyance and chemical reactions, culminating in an augmented electrochemical efficacy for hydrogen evolution.

Table 1. Equivalent circuit and impedance fit for electrochemical impedance.

Equivalent Circuit	Catalysts	R_S (Ω)	R_{CT} (Ω)
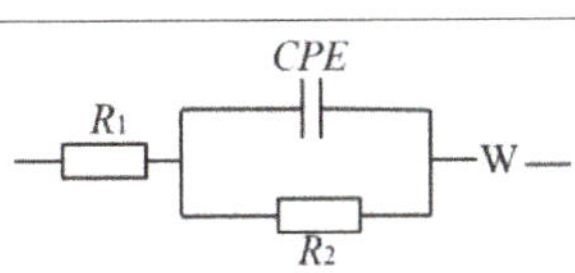	MoS_2	36.19	890.01
	Mo_2C	12.21	130.90
	Mo_2C/MoS_2-1	44.37	51.86

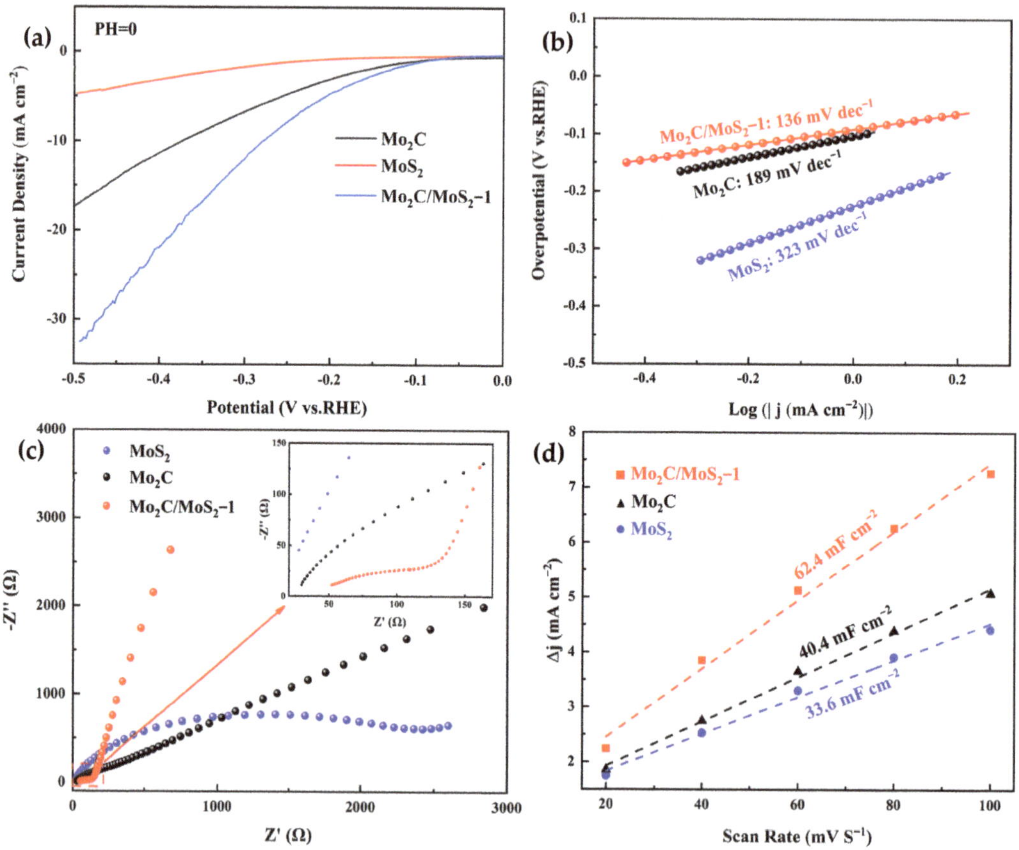

Figure 5. Electrochemical test plots of Mo_2C, MoS_2, and Mo_2C/MoS_2 composite catalysts: (**a**) LSV; (**b**) Tafel slope; (**c**) EIS; (**d**) ECSA.

Due to the direct linear relationship between the ECSA of a catalyst and the magnitude of its C_{dl}, the ECSA computation of materials can be inferred by the measurement of their C_{dl}. As depicted in Figure 5d, the ECSA of the Mo_2C/MoS_2-1 composite catalyst is discerned to be 62.4 mF cm^{-2}, which notably surpasses that of the isolated Mo_2C at 40.4 mF cm^{-2} and MoS_2 at 33.6 mF cm^{-2}. This observation indicates that the composite catalyst acquired an enlarged electrochemically active specific surface area post-synthesis, which resulted in a greater density of active sites. Consequently, this enhancement led to improved electrochemical hydrogen evolution performance, corroborating the outcomes previously illustrated by the polarization curves and Tafel slopes. An excessively high Tafel slope indicates that the Volmer reaction rate is slow, which can limit the reaction rate of the Heyrovsky reaction with the Tafel reaction, further affecting the overall conversion rate of the reaction. The ECSA of the catalyst was significantly increased after the composite of the two, which drastically reduced the Tafel slope of the catalyst and greatly accelerated the rate of electron transfer to produce H* faster. It indicates that the heterogeneous structure can form a high-speed electron transfer channel and dramatically improve the catalytic activity.

The service stability of the catalysts was tested by cyclic voltammetry and chronoamperometry, and the test results are shown in the following figures: Figure 6a shows the linear voltammetry curves of the Mo_2C/MoS_2-1 composite catalysts before and after 1000 cycles at a sweep rate of 100 mV s^{-1}, which shows that the overpotential of the catalysts increased slightly after the cycle, with little change compared with that before the cycle. Figure 6b shows the chronoamperometric current curve of the Mo_2C/MoS_2-1 composite catalyst with a starting voltage of 297 mV, and the results indicate that the current density was almost unchanged under the test condition of 12 h acidic electrolyte. The above results indicate that the Mo_2C/MoS_2-1 composite catalyst exhibited good stability under long-time endurance tests.

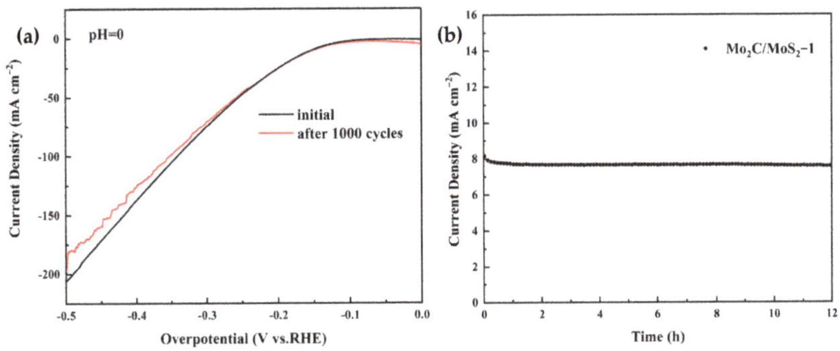

Figure 6. Stability test of Mo_2C/MoS_2-1 composite catalyst: (**a**) LSV plot after 1000 cycles of CV; (**b**) its curve plot.

3.3. Characterization of Composite Catalysts with Different Molar Ratios

Figure 7 portrays the X-ray diffraction patterns of heterogeneous catalysts comprised of variable molar proportions of Mo_2C to MoS_2. Within these spectrums, one can discern the crystalline signatures of Mo_2C at ordinal angles (2θ) of 34.4°, 38.0°, 39.4°, 52.1°, 61.5°, 69.6°, 72.4°, 74.6°, and 75.5°. These peaks correspond to the facets of the β-Mo_2C polymorph (as cataloged in PDF#35-0787) specifically at the (100), (002), (101), (102), (110), (103), (200), (112), and (201) crystallographic planes, respectively, thereby affirming the successful synthesis of Mo_2C nanoparticles via thermochemical carbonization. Moreover, at diminished Mo_2C to MoS_2 ratios, diffraction peaks were discerned at 2θ values of 13.5°, 33.1°, 35.4°, and 57.4°, which are ascribable to the (011), (200), (003), and (110) lattice planes of hexagonal MoS_2, respectively. The absence of extraneous peaks underscores the purity of the synthesized Mo_2C/MoS_2 composited catalysts across the explored molar ratios. As the molar ratio of Mo_2C to MoS_2 ascended, a concurrent diminution and broadening of MoS_2 diffraction peaks were observed, signifying the potential of the composite to attenuate interlayer distances and particle size of the lamellar MoS_2 structure. This structural alteration renders augmented accessibility to active sites, thereby enhancing the catalytic prowess of the material. Notably, at a molar ratio of 4 units of Mo_2C to MoS_2, the peaks representative of MoS_2 were not detected, attributed to the insufficiency of MoS_2 presence within the mixture.

Upon examining the XPS spectra, it is discernible that the characteristic peaks at 230.6 eV and 227.4 eV for the Mo_2C/MoS_2-1 specimen, as well as the analogous peaks for the Mo_2C/MoS_2-3 specimen depicted in Figure 8a, correspond to the Mo^{2+} $3d_{3/2}$ and Mo^{2+} $3d_{5/2}$ states of molybdenum, respectively, signifying the presence of Mo_2C. Additionally, the peaks observed at 231.9 eV and 228.0 eV for the Mo_2C/MoS_2-1 sample, along with the peaks at 232.0 eV and 228.0 eV for the Mo_2C/MoS_2-3 sample, are representative of the Mo^{4+} $3d_{3/2}$ and Mo^{2+} $3d_{5/2}$ states of molybdenum, ascribed to Mo_2C. Furthermore, the peaks at 231.9 eV and 228.0 eV about the Mo_2C/MoS_2-1 specimen and the peaks at 232.0 eV and 228.0 eV relative to the Mo_2C/MoS_2-3 specimen correspond to the Mo^{4+} $3d_{3/2}$ and Mo^{4+} $3d_{5/2}$ states, which are indicative of MoS_2. In Figure 8b, the peak positioned at 283.0 eV for both Mo_2C/MoS_2-1 and Mo_2C/MoS_2-3 is associated with C1s, denoting Mo_2C. Additionally, Figure 8c reveals that the characteristic peaks at 161.6 eV and 160.4 eV for the Mo_2C/MoS_2-1 sample and the identically characterized peaks for the Mo_2C/MoS_2-3 sample correspond to the S^{2-} $2p_{1/2}$ and S^{2-} $2p_{3/2}$ states, respectively, indicative of MoS_2, whereas the peak at 167.6 eV is suggestive of $SO_4{}^{2-}$, implicating the partial oxidation of S^{2-}

in atmospheric conditions. The aforementioned peaks are consistent across the board with no noticeable shift, implying that modifications in the molar ratio exert negligible influence on the heterojunction between Mo_2C and MoS_2. This constancy alludes to the relative stability inherent in the heterostructural composition of the Mo_2C/MoS_2 composite catalyst.

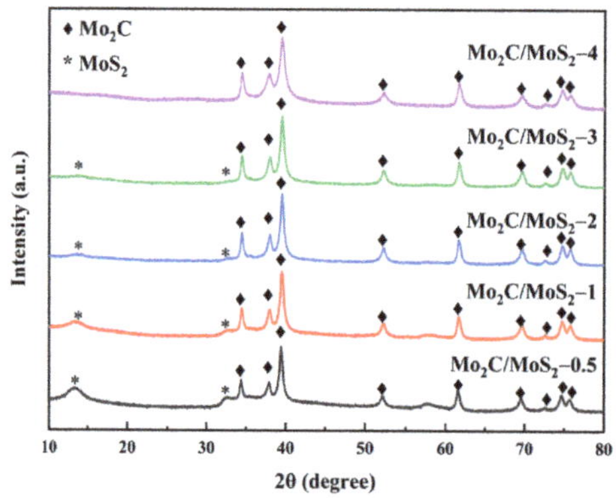

Figure 7. XRD pattern of Mo_2C/MoS_2 composite catalysts with different composite ratios.

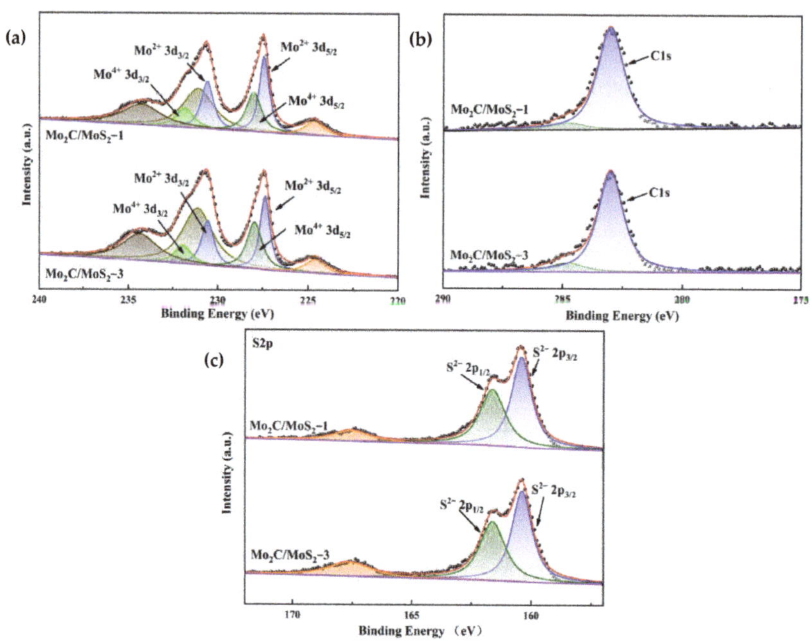

Figure 8. XPS patterns of Mo_2C/MoS_2 composite catalysts with different composite ratios: (**a**) Mo3d; (**b**) C1s; (**c**) S2p.

Figure 9 delineates the scanning electron microscopy (SEM) representations of the composite catalysts across varying molybdenum carbide to molybdenum disulfide (Mo_2C/MoS_2) molar ratios. Observations reveal a diminution in MoS_2 crystallite dimensions concomitant with an escalation in Mo_2C/MoS_2 molar proportionality. Precisely, the crystallite dimensions approximate 300 nm for Mo_2C/MoS_2-0.5; however, this metric contracts to approximately 60–100 nm for the Mo_2C/MoS_2-3 specimen, with the reduced crystallite size engendering a commensurately augmented specific surface area. Consequently, this engenders a notably superior electrochemical efficacy in hydrogen evolution reactions.

Figure 9. SEM images of Mo_2C/MoS_2 composite catalysts with different composite ratios: (**a**) Mo_2C/MoS_2-0.5; (**b**) Mo_2C/MoS_2-1; (**c**) Mo_2C/MoS_2-2; (**d**) Mo_2C/MoS_2-3; (**e**) Mo_2C/MoS_2-4.

After the integration of the two constituents, the inherently two-dimensional stratified morphology of MoS_2 underwent a transition towards a nanoflower spherical configuration, which is indicative of a considerable morphological transfiguration attributed to the influence of Mo_2C. Owing to the layered architecture, the lion's share of catalytically active sites within MoS_2 was predominantly localized at the laminae peripheries. Nonetheless, the emergent nanoflower spherical structure precipitated by the composite formation facilitated the generation of additional active cross-sections. In concert with this morphological evolution, the inherent strata manifested numerous intrinsic strain-induced defects, thereby proliferating the density of active sites conducive to catalytic processes.

3.4. Electrochemical Results of Composite Catalysts with Different Molar Ratios

Figure 10 delineates the electrocatalytic hydrogen evolution prowess of composite catalysts with varying molar ratios of Mo_2C to MoS_2, assayed in an aqueous sulfuric acid solution of 0.5 mol L^{-1}. As discerned from the linear sweep voltammetry profiles illustrated in Figure 10a, the Mo_2C/MoS_2-0.5 composite mandated the highest overpotential for achieving a current density of 10 mA cm^{-2}, in stark contrast to the Mo_2C/MoS_2-3 composite, which required the least overpotential at the equivalent current density, signifying its superior electrocatalytic hydrogen evolution capability. Corresponding to these findings, Figure 10b presents the electrocatalytic hydrogen evolution attributes of the catalyst variants stratified by their molar ratios, congruent with the linear sweep voltammetry data: an onset overpotential of 183 mV dec^{-1} for Mo_2C/MoS_2-0.5 and 150 mV dec^{-1} for Mo_2C/MoS_2-4. This suggests that a molar ratio of Mo_2C to MoS_2 that is either excessively low or high is detrimental to the electrocatalytic efficacy of the composites. The Mo_2C/MoS_2-3 composite exhibited the most negligible Tafel gradient, an indicator that this particular formulation can efficaciously mitigate the activation barrier for the electrocatalytic hydrogen evolution process, hence displaying unparalleled performance. Figure 10c exposes the electrochemical impedance spectroscopy (EIS) profiles for the composites across varying molar ratios, echoing the aforementioned empirical data: the Mo_2C/MoS_2-3 composite's Nyquist plots revealed the least pronounced curvature within the high-frequency electron transport domain, indicative of optimal electrochemical performance. The equivalent circuit diagram and specific values are shown in Table 2.

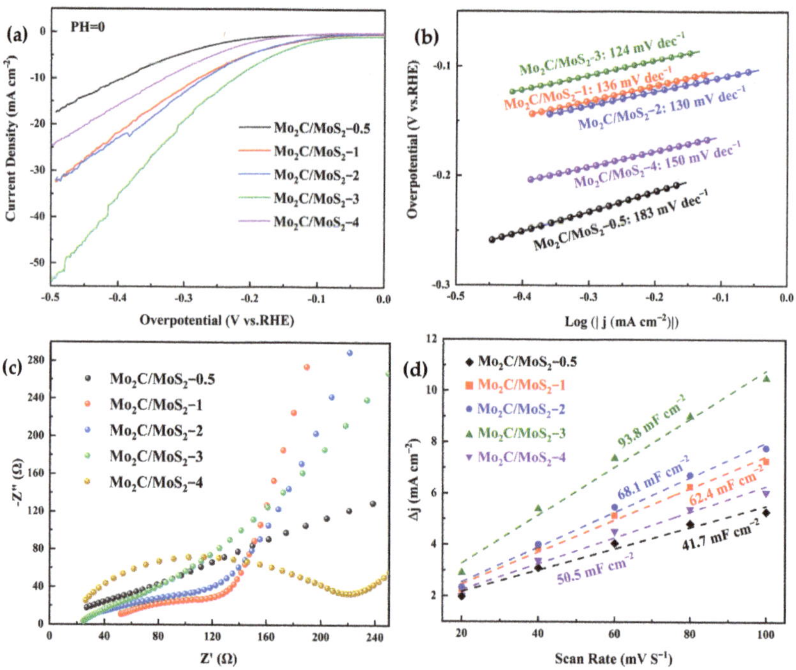

Figure 10. Electrochemical performance of Mo_2C/MoS_2 composite catalysts with different composite ratios: (**a**) LSV; (**b**) Tafel slope; (**c**) EIS; (**d**) ECSA.

Table 2. EIS values and equivalent circuit diagrams for different composite ratios.

Equivalent Circuit	Catalysts	R_S (Ω)	R_{CT} (Ω)
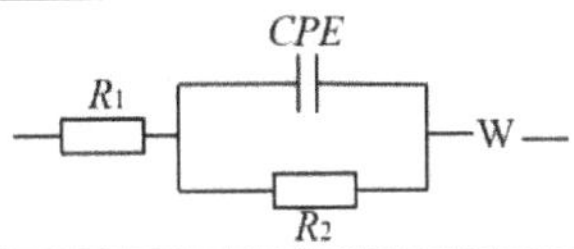	Mo_2C/MoS_2-0.5	35.23	70.28
	Mo_2C/MoS_2-1	44.37	51.86
	Mo_2C/MoS_2-2	39.45	47.86
	Mo_2C/MoS_2-3	21.67	39.20
	Mo_2C/MoS_2-4	20.25	100.37

Upon augmentation of the composite ratio, Mo_2C, serving as an electron donor, endowed MoS_2 with an enhanced concentration of electrons, thereby diminishing its electrochemical impedance. The Mo_2C/MoS_2-3 samples, with a surplus of readily available free electrons, demonstrated enhanced electrocatalytic activity and were adept at promoting synergistic electron transfer, which accelerates the reduction of protons to hydrogen. Within the array of composite catalysts with varying composite ratios, the ECSA of the catalysts was similarly deduced through the evaluation of the C_{dl} of the materials under investigation. In Figure 10d, the graphical depictions reveal that the Mo_2C/MoS_2-3 composite manifested the highest ECSA, registering a capacitance of 93.8 mF cm^{-2}. This represents a marked increase of 132% over the pristine Mo_2C catalyst and an augmentation of 179% in comparison to the unadulterated MoS_2 catalyst. As the composite ratio of Mo_2C to MoS_2 ascended, there was a discernible upward trajectory in the electrochemical active surface area of the composite catalysts. However, the ECSA of the Mo_2C/MoS_2-4 sample exhibited a subtle decline, which can be conjectured to result from an overly elevated Mo_2C to MoS_2 composite ratio. Such a ratio may diminish the synergetic interaction between the two constituents, predominantly relegating Mo_2C to serve as the principal component of the catalyst.

Mo_2C is a narrow-band semiconductor material with a forbidden bandwidth of 1.26 eV, and MoS_2 has a forbidden bandwidth of 2.31 eV, and a type-II heterostructure will be formed after compositing the two. Due to the difference in electrical conductivity between Mo_2C and MoS_2, the Mo_2C/MoS_2 composite catalyst will form a heterostructure similar to a P-N junction at the composite interface [30,31]. When the PN junction-type heterojunction is formed, the electrons close to the P-N interface will diffuse to the P region, while the holes close to the P-N interface will diffuse to the N region, leaving the negatively charged P region ions. Due to the high concentration of negatively and positively charged ions assembled at the semiconductor interface, a built-in electric field will be formed, which can drive the electrons and holes to be transported in the opposite direction, greatly enhancing the charge separation efficiency. For HER, the strong interfacial electric field formed by the PN junction-type heterojunction can significantly increase the d-band center of the catalyst, enhance the adsorption of H*, and contribute to the diffusion of metal ions. A comparison of the results of this work with those of other researchers is shown in Table 3

Figure 11 shows the side view of the stability of the Mo_2C/MoS_2-3 catalysts. It can be seen from Figure 11a that the cyclic voltammetry test was carried out at a sweep rate of 100 mV s^{-1}, and after cycling for 1000 laps, the LSV curves of the Mo_2C/MoS_2-3 catalysts showed a relatively small change, and the overpotentials slightly increased at the same current density, which indicated a slight decrease in the catalytic activity of the catalysts. Figure 11b shows that it curve of the Mo_2C/MoS_2-3 catalyst showed no obvious fluctuation after 12 h of the timed current test at an initial spot level of 223 mV, which exhibited good stability. The above results indicate that the Mo_2C/MoS_2-3 catalysts have excellent stability in long-time electrochemical reactions.

Table 3. Comparison of the results of this thesis with the experimental results of other researchers.

Catalysts	Over Potential at 10 mA/cm^2 (mV)	Tafel Slope (mV/dec)	ECSA (mF/cm^2)	R_{CT} (Ω)	References
Mo_2C/MoS_2-3	223.4	124	93.8	39.2	This work
Mo_2C&MoS_2@NSC_3	220	85.5	5.6	/	[32]
Mo_2C/MoS_2-25 min	63	53	146	7.5	[33]
NC@Mo_2C@MoS_2-(Ni)	205	61.4	4.7	45	[34]
3D MoS_2@Mo_2C NS	67	37	64.2	/	[35]

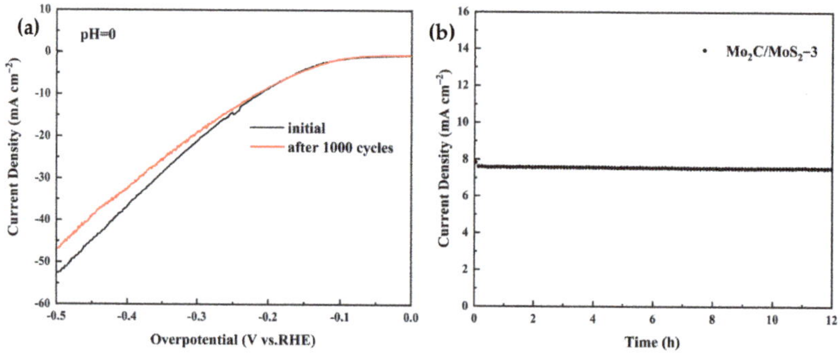

Figure 11. Stability test of Mo_2C/MoS_2-3 composite catalyst: (**a**) LSV plot after 1000 cycles of CV; (**b**) its curve plot.

4. Conclusions

In summation, Mo_2C nanoparticles were efficaciously synthesized through a procedure of high-temperature carbonization, subsequently leading to the creation of Mo_2C/MoS_2 composite catalysts via the hydrothermal method. Characterization of these composites utilizing advanced material characterization methodologies, such as X-ray diffraction (XRD), scanning electron microscopy (SEM), transmission electron microscopy (TEM), and X-ray photoelectron spectroscopy (XPS), has unequivocally confirmed the successful integration of Mo_2C into a heterogenous structural amalgamation with MoS_2. This fusion has substantially curtailed the proclivity of MoS_2 to aggregate, thereby engendering a prolific increase in defects within the MoS_2 crystals and unveiling a greater quantity of catalytically active sites. The potential for further enhancement of the composite catalysts' electronic structure lies in the modulation of the constituent ratios of Mo_2C to MoS_2, which precipitates enhanced kinetics in the adsorption, desorption, and reduction processes of H^+ ions. When juxtaposed with the singular Mo_2C and MoS_2 crystals, it is evident that the catalytic efficacy of the composite is markedly superior. Electrochemical assessments, specifically hydrogen evolution reaction (HER) studies, demonstrated that the Mo_2C/MoS_2-3 sample heralded an overpotential of merely 224 mV and a Tafel slope of 124 mV dec^{-1} at a current density of 10 mA cm^{-2}, showcasing exceptional HER performance. This remarkable performance is ascribable to the heterostructures interlinking the composite components, where Mo_2C affords electron conduits by its noble metal-like characteristics, thereby ameliorating the electrical conductivity of MoS_2. Concurrently, MoS_2's stratified structure has a wealth of crystal defects and an expansive specific surface area that avails additional active sites germane to the catalyst's function. The empirical data coalesce to indicate that the intrinsic heterogeneous structure of the Mo_2C/MoS_2 composite catalyst acts to significantly bolster its catalytic prowess.

Author Contributions: Conceptualization, C.Z.; Methodology, C.Z.; Validation, C.Z.; Formal analysis, C.Z.; Resources, Z.P. and Y.T.; Data curation, C.Z.; Writing—original draft, C.Z.; Writing—review & editing, Z.P. and Y.T.; Supervision, Y.T.; Project administration, Y.T.; Funding acquisition, Z.P. and Y.T. All authors have read and agreed to the published version of the manuscript.

Funding: This work was supported by the priority academic program development of Jiangsu Higher Education Institution (PAPD).

Institutional Review Board Statement: Not applicable.

Informed Consent Statement: Not applicable.

Data Availability Statement: The data presented in this study are available on request from the corresponding author.

Acknowledgments: The authors gratefully acknowledge the assistance from Zhigang Pan and Yaqiu Tao from NJTECH and the staff from the State Key Laboratory of Materials-Oriented Chemical Engineering.

Conflicts of Interest: The authors declare no conflicts of interest.

References

1. Turner, J.A. Sustainable hydrogen production. *Science* **2004**, *305*, 972–974. [CrossRef] [PubMed]
2. Shoaib, M.; Naz, M.Y.; Shukrullah, S.; Munir, M.A.; Irfan, M.; Rahman, S.; Ghanim, A.A.J. Dual S-Scheme Heterojunction $CdS/TiO_2/g\text{-}C_3N_4$ Photocatalyst for Hydrogen Production and Dye Degradation Applications. *ACS Omega* **2023**, *8*, 43139–43150. [CrossRef] [PubMed]
3. Sun, F.; Song, J.; Wen, H.; Cao, X.; Zhao, F.; Qin, J.; Mao, W.; Tang, X.; Dong, L.; Long, Y. Ce^{4+}/Ce^{3+} Redox Effect-Promoted CdS/CeO_2 Heterojunction Photocatalyst for the Atom Economic Synthesis of Imines under Visible Light. *Inorg. Chem.* **2023**, *62*, 17961–17971. [CrossRef]
4. Lu, Z.; Yang, H.; Liu, Q.; Luo, J.; Feng, L.; Chu, L.; Liu, X. Nb_2AlC MAX Nanosheets Supported Ru Nanocrystals as Efficient Catalysts for Boosting pH-Universal Hydrogen Production. *Small* **2023**, *20*, e2305434. [CrossRef]
5. Morales-Guio, C.G.; Stern, L.A.; Hu, X.L. Nanostructured hydrotreating catalysts for electrochemical hydrogen evolution. *Chem. Soc. Rev.* **2014**, *43*, 6555–6569. [CrossRef]
6. Popczun, E.J.; McKone, J.R.; Read, C.G.; Biacchi, A.J.; Wiltrout, A.M.; Lewis, N.S.; Schaak, R.E. Nanostructured Nickel Phosphide as an Electrocatalyst for the Hydrogen Evolution Reaction. *J. Am. Chem. Soc.* **2013**, *135*, 9267–9270. [CrossRef] [PubMed]
7. Alptekin, F.M.; Celiktas, M.S. Review on Catalytic Biomass Gasification for Hydrogen Production as a Sustainable Energy Form and Social, Technological, Economic, Environmental, and Political Analysis of Catalysts. *ACS Omega* **2022**, *7*, 24918–24941. [CrossRef] [PubMed]
8. Anniwaer, A.; Chaihad, N.; Zhang, M.; Wang, C.; Yu, T.; Kasai, Y.; Abudula, A.; Guan, G. Hydrogen-rich gas production from steam co gasification of banana peel with agricultural residues and woody biomass. *Waste Manag.* **2021**, *125*, 204–214. [CrossRef]
9. Luo, Y.; Chen, J.; Wang, T. Evaluation of the Effect of CaO on Hydrogen Production by Sorption-Enhanced Steam Methane Reforming. *ACS Omega* **2024**, *9*, 5330–5337. [CrossRef] [PubMed]
10. Yusuf, B.O.; Umar, M.; Kotob, E.; Abdulhakam, A.; Taialla, O.A.; Awad, M.M.; Hussain, I.; Alhooshani, K.R.; Ganiyu, S.A. Recent Advances in Bimetallic Catalysts for Methane Steam Reforming in Hydrogen Production: Current Trends, Challenges, and Future Prospects. *Chem. Asian J.* **2023**. *online version of record*. [CrossRef] [PubMed]
11. Maček Kržmanc, M.; Daneu, N.; Čontala, A.; Santra, S.; Kamal, K.M.; Likozar, B.; Spreitzer, M. $SrTiO_3/Bi_4Ti_3O_{12}$ Nanoheterostructural Platelets Synthesized by Topotactic Epitaxy as Effective Noble-Metal-Free Photocatalysts for pH-Neutral Hydrogen Evolution. *ACS Appl. Mater. Interfaces* **2021**, *13*, 370–381. [CrossRef] [PubMed]
12. Yang, Y.; Liu, L.; Wu, J.; Qian, X. Hierarchical $Bi_2S_3@CoS_2/MoS_2$ core-shelled microspheres as bifunctional electrode catalysts for efficient triiodide reduction and alkaline hydrogen evolution reactions. *J. Alloys Compd.* **2022**, *924*, 166548. [CrossRef]
13. Li, L.; Wang, P.; Shao, Q.; Huang, X. Metallic nanostructures with low dimensionality for electrochemical water splitting. *Chem. Soc. Rev.* **2020**, *49*, 3072–3106. [CrossRef] [PubMed]
14. Fernandes, V.R.; Pinto, A.M.F.R.; Rangel, C.M. Hydrogen production from sodium borohydride in methanol–water mixtures. *Int. J. Hydrogen Energy* **2010**, *35*, 9862–9868. [CrossRef]
15. Merki, D.; Hu, X. Recent developments of molybdenum and tungsten sulfides as hydrogen evolution catalysts. *Energy Environ. Sci.* **2011**, *4*, 3878–3888. [CrossRef]
16. Wu, H.; Shi, L.; Feng, C.; Ding, Y. Synthesis of $MoS_2/CoSe_2$-x hybrids as electrocatalysts for hydrogen evolution reaction. *Ionics* **2022**, *28*, 1337–1345. [CrossRef]
17. Bernsmeier, D.; Selve, S.; Nissen, J.; Maticiuc, N.; Ibaceta-Jana, J.; Szyszka, B.; Muydinov, R. Gas Flow Sputtering of Pt/C Films and their Performance in Electrocatalytic Hydrogen Evolution Reaction. *ChemPhysChem* **2023**, *24*, e202200650. [CrossRef]
18. Liu, Y.; Xu, S.; Zheng, X.; Lu, Y.; Li, D.; Jiang, D. Ru-doping modulated cobalt phosphide nanoarrays as efficient electrocatalyst for hydrogen evolution rection. *J. Colloid Interface Sci.* **2022**, *625*, 457–465. [CrossRef] [PubMed]

19. Park, S.H.; Jo, T.H.; Lee, M.H.; Kawashima, K.; Mullins, C.B.; Lim, H.-K.; Youn, D.H. Highly active and stable nickel–molybdenum nitride (Ni_2Mo_3N) electrocatalyst for hydrogen evolution. *J. Mater. Chem. A* **2021**, *9*, 4945–4951. [CrossRef]
20. Liu, Y.; Kelly, T.G.; Chen, J.G.; Mustain, W.E. Metal Carbides as Alternative Electrocatalyst Supports. *Acs Catal.* **2013**, *3*, 1184–1194. [CrossRef] [PubMed]
21. Zhong, Y.; Zhao, X.J.; Feng, Y.L.; Lu, H.; Yin, P.; Chen, Z.R.; Jie, Y.; Guo, J.Y.; Pu, M.; Yan, H. DFT study on the electrochemical synthesis of ammonia over $Mo_2C(121)$ with N-doping. *Mol. Catal.* **2022**, *530*, 112637. [CrossRef]
22. Noerskov, J.K.; Bligaard, T.; Logadottir, A.; Kitchin, J.R.; Chen, J.G.; Pandelov, S.; Stimming, U. Trends in the Exchange Current for Hydrogen Evolution. *ChemInform* **2005**, *36*, 23–26. [CrossRef]
23. Li, H.; Li, Y.; Aljarb, A.; Shi, Y.; Li, L.-J. Epitaxial Growth of Two-Dimensional Layered Transition-Metal Dichalcogenides: Growth Mechanism, Controllability, and Scalability. *Chem. Rev.* **2018**, *118*, 6134–6150. [CrossRef] [PubMed]
24. Jellinek, F.; Brauer, G.; MÜLler, H. Molybdenum and Niobium Sulphides. *Nature* **1960**, *185*, 376–377. [CrossRef]
25. Zhou, D.; Shu, H.-b.; Hu, C.; Jiang, L.; Liang, P.; Chen, X. Unveiling the Growth Mechanism of MoS_2 with Chemical Vapor Deposition: From Two-Dimensional Planar Nucleation to Self-Seeding Nucleation. *Cryst. Growth Des.* **2018**, *18*, 1012–1019. [CrossRef]
26. Ma, W.; Dong, Y.; Li, J.; Wang, Y.; Wang, T.; Zheng, X.; Deng, Y. Recent strategies for improving the catalytic activity of ultrathin transition metal sulfide nanosheets toward the oxygen evolution reaction. *Mater. Today Energy* **2024**, *40*, 101492. [CrossRef]
27. Zahra, R.; Pervaiz, E.; Yang, M.H.; Rabi, O.; Saleem, Z.; Ali, M.; Farrukh, S. A review on nickel cobalt sulphide and their hybrids: Earth abundant, pH stable electro-catalyst for hydrogen evolution reaction. *Int. J. Hydrogen Energy* **2020**, *45*, 24518–24543. [CrossRef]
28. Bagus, P.S.; Ilton, E.S.; Nelin, C.J. The interpretation of XPS spectra: Insights into materials properties. *Surf. Sci. Rep.* **2013**, *68*, 273–304. [CrossRef]
29. Li, C.; Zhang, N.; Gao, P. Lessons learned: How to report XPS data incorrectly about lead-halide perovskites. *Mater. Chem. Front.* **2023**, *7*, 3797–3802. [CrossRef]
30. Deng, R.X.; Zhang, H.R.; Zhang, Y.H.; Chen, Z.Y.; Sui, Y.P.; Ge, X.M.; Liang, Y.J.; Hu, S.K.; Yu, G.H.; Jiang, D. Graphene/Mo_2C heterostructure directly grown by chemical vapor deposition. *Chin. Phys. B* **2017**, *26*, 67901. [CrossRef]
31. Du, H.; Liu, Y.-N.; Shen, C.-C.; Xu, A.-W. Nanoheterostructured photocatalysts for improving photocatalytic hydrogen production. *Chin. J. Catal.* **2017**, *38*, 1295–1306. [CrossRef]
32. Wang, Q.; Yu, R.; Shen, D.; Liu, G.; Hong Luo, K.; Wu, C.; Gu, S. One-pot synthesis of Mo_2C&MoS_2 loaded on N/S co-doped carbon materials as the electrocatalyts for hydrogen evolution reaction. *Fuel* **2022**, *318*, 123615. [CrossRef]
33. Zhao, Z.; Qin, F.; Kasiraju, S.; Xie, L.; Alam, M.K.; Chen, S.; Wang, D.; Ren, Z.; Wang, Z.; Grabow, L.C.; et al. Vertically Aligned MoS_2/Mo_2C hybrid Nanosheets Grown on Carbon Paper for Efficient Electrocatalytic Hydrogen Evolution. *Acs Catal.* **2017**, *7*, 7312–7318. [CrossRef]
34. Gong, L.; Mu, X.; Li, Q.; Ma, L.; Xiong, Y.C.; Li, R. Rational design of Ni-induced NC @Mo_2C@MoS_2 sphere electrocatalyst for efficient hydrogen evolution reaction in acidic and alkaline media. *Int. J. Hydrogen Energy* **2021**, *46*, 5250–5258. [CrossRef]
35. Yang, S.; Wang, Y.; Zhang, H.; Zhang, Y.; Liu, L.; Fang, L.; Yang, X.; Gu, X.; Wang, Y. Unique three-dimensional Mo_2C@MoS_2 heterojunction nanostructure with S vacancies as outstanding all-pH range electrocatalyst for hydrogen evolution. *J. Catal.* **2019**, *371*, 20–26. [CrossRef]

materials

MDPI

Article

Single-Atom Underpotential Deposition at Specific Sites of N-Doped Graphene for Hydrogen Evolution Reaction Electrocatalysis

Haofei Wu [1,2,3], Qiwen Zhang [1,2,3], Shufen Chu [4], Hao Du [3,5], Yanyue Wang [3,5] and Pan Liu [1,2,3,*]

1 State Key Laboratory of Metal Matrix Composites, School of Materials Science and Engineering, Shanghai Jiao Tong University, Shanghai 200240, China; sjtumsewu@sjtu.edu.cn (H.W.)
2 Shanghai Key Laboratory of Advanced High-Temperature Materials and Precision Forming, Shanghai Jiao Tong University, Shanghai 200240, China
3 Shanghai Jiao Tong University—JA Solar New Energy Materials Joint Research Center, Shanghai 200240, China
4 National Engineering Research Center of Light Alloy Net Forming and State Key Laboratory of Metal Matrix Composite, Shanghai Jiao Tong University, Shanghai 200240, China
5 JA Solar Technology Co., Ltd., Beijing 100160, China
* Correspondence: panliu@sjtu.edu.cn

Abstract: Single-atom catalysts (SACs) have the advantages of good active site uniformity, high atom utilization, and high catalytic activity. However, the study of its controllable synthesis still needs to be thoroughly investigated. In this paper, we deposited Cu SAs on nanoporous N-doped graphene by underpotential deposition and further obtained a Pt SAC by a galvanic process. Electrochemical and spectroscopic analyses showed that the pyridine-like N defect sites are the specific sites for the underpotential-deposited SAs. The obtained Pt SAC exhibits a good activity in a hydrogen evolution reaction with a turnover frequency of 25.1 s^{-1}. This work reveals the specific sites of UPD of SAs on N-doped graphene and their potential applications in HERs, which provides a new idea for the design and synthesis of SACs.

Keywords: underpotential deposition; single-atom catalyst; hydrogen evolution reaction; N-doped graphene

1. Introduction

Worldwide interest in H_2 as a clean chemical energy source to replace fossil fuels and avoid CO_2 emissions is increasing [1–3]. The electrocatalytic hydrogen evolution reaction (HER) based on renewable electricity is a promising, safe, scalable, low-cost, green method of producing H_2 [4–6]. The noble metal Pt is considered the most efficient catalyst for use in the HER [7–9]. However, owing to its high cost, further consideration is required to maximize the efficiency of precious metal utilization and improve the activity of the catalyst by modulating its electronic structure [10–12].

Carbon is widely used as a multifunctional material in numerous fields. In the field of hydrogen energy, due to its good electrical conductivity, high chemical stability, and high specific surface area, it can not only be used as a hydrogen storage material [13,14] but can also widely be used as a substrate material for loaded catalysts. Loaded nanocatalysts are immobilized on supports, and the electronic structures of the active sites of the catalysts can be efficiently regulated via strong metal–support interactions, which are described as electronic metal–support interactions (EMSIs) [15–17]. Single-atom catalysts (SACs) minimize the catalytic structures with well-defined active sites. Their homogeneous atomic coordination environments render them as ideal simplified model systems for use in studying catalytic reaction mechanisms [18–20]. Strong EMSIs not only stabilize single-atom metals by forming thermodynamically favorable metal–support bonds but also modulate the electronic structures of single-atom metals and their catalytic activities via electron transfer, leading to charge redistribution [21–23].

Now, commonly used synthesis methods for SACs often require the usage of high energies to drive the formation of SAs. For example, Shah et al. synthesized a Co SAC on RuO_2 spheres by hydrothermal reactions coupled with calcination for HER and OER (oxygen evolution reaction) bifunctional catalysis [24]. Jin et al. prepared K_2PtCl_4@NC-M SAC using high-energy ball milling at 400 rpm for 20 min, which showed a 17-fold enhanced mass activity over commercial catalysts [25]. Qu et al. used defects in defective graphene (DG) to trap Pt atoms at high temperatures and gradually converted bulk Pt into Pt Sas; the prepared Pt SACs/DG exhibited high HER activity [26]. Underpotential deposition is a mild typical surface self-limiting reaction that has been used for the growth of metal single-atom layers [27]. Since UPD tends to obtain continuous layers due to the uniform distribution of the substrate surface work function, Shi et al. proposed that single layers can be further upgraded to single atoms (SAs) by using materials with non-uniformly distributed Lewis base sites as substrates for underpotential deposition [28]. Transition metal disulfide compounds and carbon-based materials have been shown to be supports that can be used for the underpotential deposition of single atoms of copper, but the underpotential deposition-specific sites have been considered to be the S, Se, and O sites in previous studies, and studies on N sites as underpotential deposition sites are not yet understood [28–30].

In this work, we prepared a Pt SAC on nanoporous nitrogen-doped graphene (np-NG/Pt) by UPD and a galvanic process. Compared to conventional methods, the preparation of single atoms using UPD is carried out under milder conditions and is more energy efficient. Detailed spectroscopic and electrochemical data indicate that among the different kinds of N defect sites in N-doped graphene, the pyridine-like nitrogen defects are the specific sites for the formation of Cu SAs by UPD. The Pt SACs further formed by the galvanic process replacing the Cu SAs show good activity in an acidic HER.

2. Materials and Methods

2.1. Synthesis of Nanoporous Nitrogen-Doped Graphene (np-NG)

The fabrication of np-NG has been reported in detail in previous works [31,32]. Firstly, a $Ni_{30}Mn_{70}$ (atomic ratio) ingot, obtained by an arc-melting method, was cold-rolled into $Ni_{30}Mn_{70}$ sheets with a thickness of 150 μm. The $Ni_{30}Mn_{70}$ sheets were chemically dealloyed in a 1.0 M $(NH_4)_2SO_4$ aqueous solution at 50 °C for 5 h, then thoroughly rinsed with deionized water and ethanol and dried in a vacuum oven at room temperature to obtain nanoporous Ni templates.

The np-Ni template was loaded into a quartz tube (φ50 × φ50 × 1400 mm) of a three-heating zone furnace, and np-NG was fabricated on the np-Ni template by a chemical vapor deposition (CVD) process. Melamine was used as the carbon and nitrogen sources for growing np-NG. A total of ~0.25 g of melamine powder and the np-Ni sheet were loaded into porcelain boats and placed in the central heating zone and the adjacent downstream heating zone, respectively. The downstream heating zone was first heated from room temperature to 800 °C within 30 min to anneal nanoporous Ni under the mixed gas atmosphere of Ar (200 sccm) and H2 (200 sccm). Meanwhile, the central heating zone was gradually ramped to 200 °C within 30 min. During the 5 min pre-reduction annealing, melamine powders were further heated to 300 °C. When the temperature of the central heating zone approaches 300 °C, melamine powders start to evaporate and form a continuous graphene film on the Ni substrate. The furnace was opened immediately after 7 min of graphene growth and cooled to room temperature using a fan to obtain np-Ni/NG. The samples without N doping were prepared in a similar way as np-Ni/NG, with the gas atmosphere and annealing temperature unchanged, and the precursor was changed to toluene. When np-Ni was pre-reduced and annealed for 5 min, toluene gas was passed through the furnace and the reaction was carried out for 5 min at 800 °C, then the furnace was opened immediately and cooled down to room temperature using a fan to obtain the np-Ni/G.

The np-Ni/NG and np-Ni/G sheets were etched with a 2.0 M HCl solution containing 5 mM $FeCl_3$ at 50 °C overnight to completely etch Ni and then rinsed five times with deionized water. Then, N-doped np-NG and non-N-doped np-G can be obtained by the freeze-drying technique. According to XPS, the N content is about 6.3%.

2.2. Synthesis of Cu SAC and Pt SAC

The Pt SAC was synthesized according to the previous report [22]. The np-NG was mixed with a 0.5% Nafion ethanol solution and sonicated to an ink with a concentration of 4 mg/mL. The φ5 mm glassy carbon (GC) electrodes were polished using alumina powder, cleaned in ethanol for a short time by ultrasonication, and dried at room temperature. A total of 5 μL of np-NG ink was dropped onto the glassy carbon electrode and dried under a hot lamp to make a working electrode. The preparation of single atoms on np-NG used an underpotential deposition (UPD) method, using a Ag/AgCl electrode (3.0 M KCl, E^θ = 0.21 V) as a reference electrode and a 1 × 1 cm^2 Pt sheet as a counter electrode.

Firstly, Cu atoms were underpotentially deposited on the np-NG by controlling the potential at +0.10 V vs. Ag/AgCl for 120 s in a 0.1 M H_2SO_4 saturated Ar solution containing 2 mM $CuSO_4$, and this sample was named np-NG/Cu. The np-NG/Cu sample was washed with deionized water and then transferred to an Ar-saturated solution of 50 mM H_2SO_4 containing 2 mM K_2PtCl_4 and allowed to stand for 10 min at open circuit potential to displace the Cu with Pt (II). The new sample was termed as np-NG/Pt.

2.3. Structure Characterizations

The morphologies of different samples were characterized by scanning electron microscope (SEM, 3D Versa, FEI, Hillsboro, OR, USA) at a 20 kV beam voltage. The chemical compositions of the samples were investigated by X-ray photoelectron spectroscopy (XPS) utilizing the AXIS UltraDLD XPS system (Kratos, Shimadzu, Kyoto, Japan). High-resolution TEM (HRTEM) images of all samples were obtained by JEOL ARM 200F (Tokyo, Japan) equipped with a spherical aberration corrector for the image-forming objective lens operating at a 200 kV acceleration voltage. The high-angle annular dark-field scanning transmission electron microscopy (HAADF-STEM) images were obtained by JEOL ARM 200F equipped with a spherical aberration corrector for the probe-forming lens and an energy-dispersive X-ray spectrometer (EDS). X-ray absorption fine structure (XAFS) measurements were conducted on the BL11B beamline at Shanghai Synchrotron Radiation Facility (SSRF), Shanghai, China.

2.4. Electrochemical Test

The electrochemical performance was tested in a three-electrode system with a 0.5 M H_2SO_4 electrolyte on a CHI 760E electrochemical workshop. A graphite electrode was used as the counter electrode and the Ag/AgCl electrode (3.0 M KCl, E^θ = 0.21 V) was used as the reference electrode. Different catalysts were loaded on the φ5 mm GC electrode as the working electrode. The loading amount of commercial 20% Pt/C or different np-NG catalysts were all 100 μg/cm^2. The linear sweep voltammetry (LSV) curve was obtained at a scan rate of 5 mV/s with a 90% iR drop correction. EIS was carried out in the frequency range of 100 kHz to 0.01 Hz at the overpotentials of 50 mV in the 0.5 M H_2SO_4 electrolyte. Relative standard errors for all components were controlled to <10% in the EIS fitting.

Electrochemical active surface areas (ECSAs) of different Pt catalysts were obtained by calculating H underpotential adsorption (H_{UPD}) by cyclic voltammetry tests in the range of −0.21 V~0.5 V (vs. Ag/AgCl) in the 0.5 M H_2SO_4 electrolyte. The equation for calculating the *ECSA* of *Pt* is shown in Equation (1).

$$ECSA = \frac{S_H / v}{0.21(\mathrm{mC/cm^2}) \cdot m_{Pt}} \tag{1}$$

where S_H is the integral area of the H_{UPD} peak, v is the sweep speed, and m_{Pt} is the Pt loading mass.

2.5. Turnover Frequency (TOF) Calculations for the HER

The TOF calculation starts with the number of moles of active sites of the catalyst, n. For SACs, one atom can be considered as one active site, so n can be obtained from Equation (2).

$$n = \frac{m}{M} \tag{2}$$

where m is the loading mass of a single atom and M is the molar mass of a single atomic element. For Pt nanoparticles, the number of catalytically active sites can be deduced from the ECSA since it has been obtained by H underpotential deposition; the n can be obtained from Equation (3).

$$n = \frac{\rho \cdot ECSA \cdot m_{Pt}}{N_A} \tag{3}$$

where $ECSA$ is the electrochemical active surface area; N_A is the Avogadro constant ($\approx 6.022 \times 10^{23}$); m_{Pt} is the Pt loading mass; and ρ is the number of active sites in a plane per unit area of Pt (1.637×10^{15} active atoms/cm^2) [33,34].

The TOF value is calculated based on Equation (4).

$$TOF = \frac{J \cdot A}{2 \cdot F \cdot n} \tag{4}$$

where A is the geometric area of the φ5 GC electrode (0.196 cm^2); F is the Faraday constant (96,485 C/mol); n is the number of moles of active materials that are loaded onto the electrode; and J is the is the current density. J can be categorized into two types; one is the exchange current density (J_0) [35] and the other is the kinetic current density (J_k) [36]. For the HER, which is less affected by the mass transfer, the measured current can be approximated to be the kinetic current, so here we choose to use the more intuitive J_k to calculate the TOF.

3. Results and Discussion

Figure 1a shows a schematic of the UPD of single atoms. The np-NG was obtained by using nanoporous Ni as a template, and its biconnected porous structure enabled better self-support, a larger specific surface area, and better mass transfer channels as a reaction support (Figure S1). Compared to 2D graphene, the high curvature introduces more defects for np-NG that can be used as growth sites for single atoms, as shown in the aberration-corrected high-angle annular dark-field scanning transmission electron microscopy (HAADF-STEM) image in Figure 1b. The blue hexagon is a typical graphene six-membered ring structure and the yellow area is a defect that lacks a carbon atom and can be used as the location to hold metal atoms. To enable the self-terminating growth of single-atom Pt on the np-NG carriers, Cu atoms are first underpotentially deposited on np-NG (denoted np-NG/Cu, Figure 1c). The Cu atoms are then galvanically exchanged with Pt(II) to form the np-NG-supported single-atom Pt samples (denoted np-NG/Pt, Figure 1d). HAADF-STEM images show atomically dispersed metal atoms (bright spots) on the np-NG/Cu and np-NG/Pt. Surface-confined UPD enables the formation of energetically favorable metal–support bonds and automatically terminates the sequential formation of metal bonds [28]. Energy-dispersive X-ray spectroscopy elemental mapping reveals that the Cu and Pt atoms are homogeneously dispersed (Figure 1e,f).

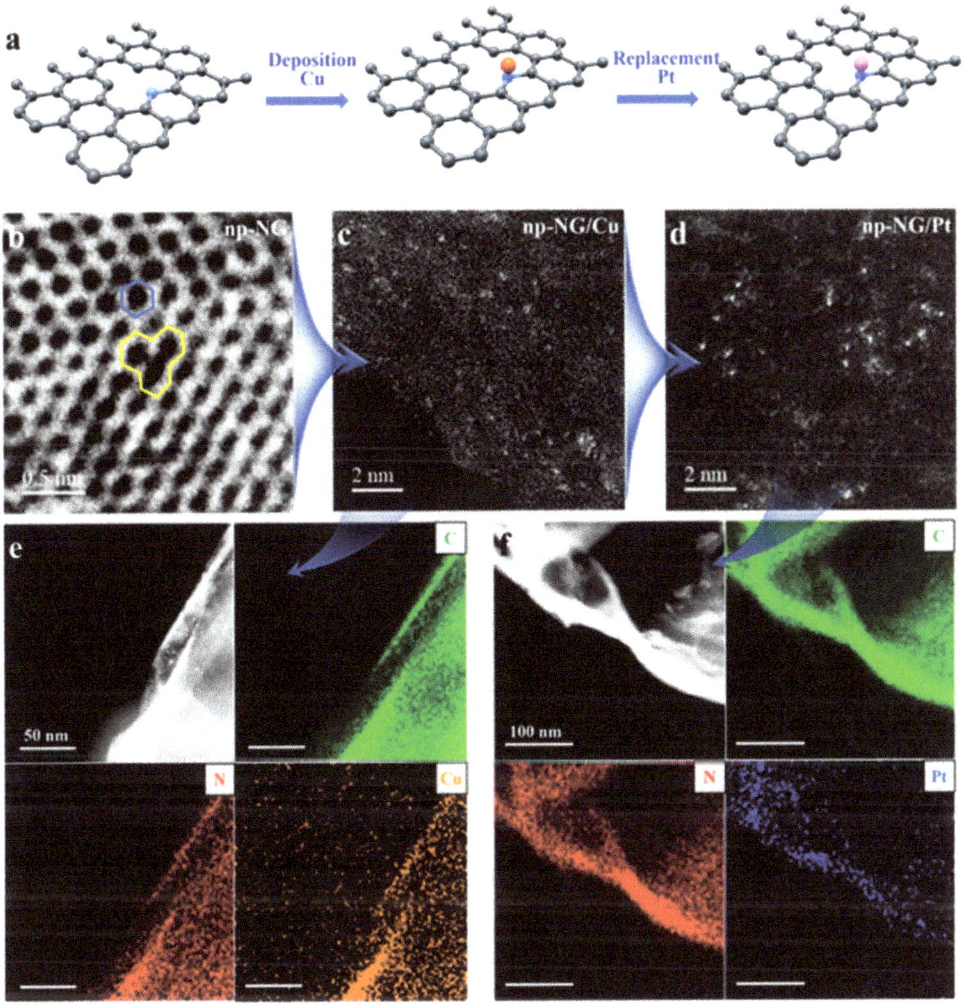

Figure 1. Single-atom design and characterization of the np-NG support. (**a**) The schematic of the UPD of single atoms. (**b**) HAADF-STEM image of defective carbon hexagon structure of np-NG. HAADF-STEM images of (**c**) np-NG/Cu and (**d**) np-NG/Pt. Corresponding elemental mapping of (**e**) np-NG/Cu and (**f**) np-NG/Pt.

The redox peaks of Cu within the different samples were compared within a wide potential window (−0.4~0 V) to exclude the influences of other factors. All the CV curves were obtained by sweeping from positive to negative potentials (cathodic sweep) and then from negative to positive potentials (anodic sweep). In the Cu foil sample, the reduction peak b is present irrespective of the presence of $CuSO_4$, which may be due to the oxide layer on the surface [37]. However, it can be seen that in the electrolyte containing $CuSO_4$, the intensity of the b peak is enhanced due to the superimposed reduction current of Cu^{2+}. The Cu^{2+} reduction current occurs at a position between −0.1 and ~0 V, representing the bulk deposition peak b of Cu (Figure 2a1) [38]. In the cyclic voltammograms of the clean glassy carbon electrode (GC) and the GC with a drop of Nafion solution, only the bulk deposition peaks of Cu appear (Figure S2). No significant reduction reaction occurred within the 0~0.5 V voltage window (Figure 2a2). In contrast, in the cyclic voltammograms of the graphene samples, in addition to the bulk deposition peaks of Cu, reduction peaks are observed at more positive deposition potentials, suggesting strong interactions between Cu and graphene. However, in the cyclic voltammograms of graphene with and without N defects, a difference in the UPD of Cu is observed in Figure 2b1,c1. The intensities of the reduction peaks of UPD and the oxidation peaks upon dissolution when N defects are present (Figure 2c2, $a_{np\text{-}NG}'$ peak, 0.35 mA/cm^2@0.22 V) are significantly stronger than those of the np-G samples without N defects (Figure 2b2, $a_{np\text{-}G}'$ peak, 0.16 mA/cm^2@0.22 V). The presence of a weak UPD peak in the curve of np-G may be caused by O defects introduced during the etching of the Ni template. Therefore, N defects are considered as specific sites for the UPD of single Cu atoms on nanoporous graphene substrates.

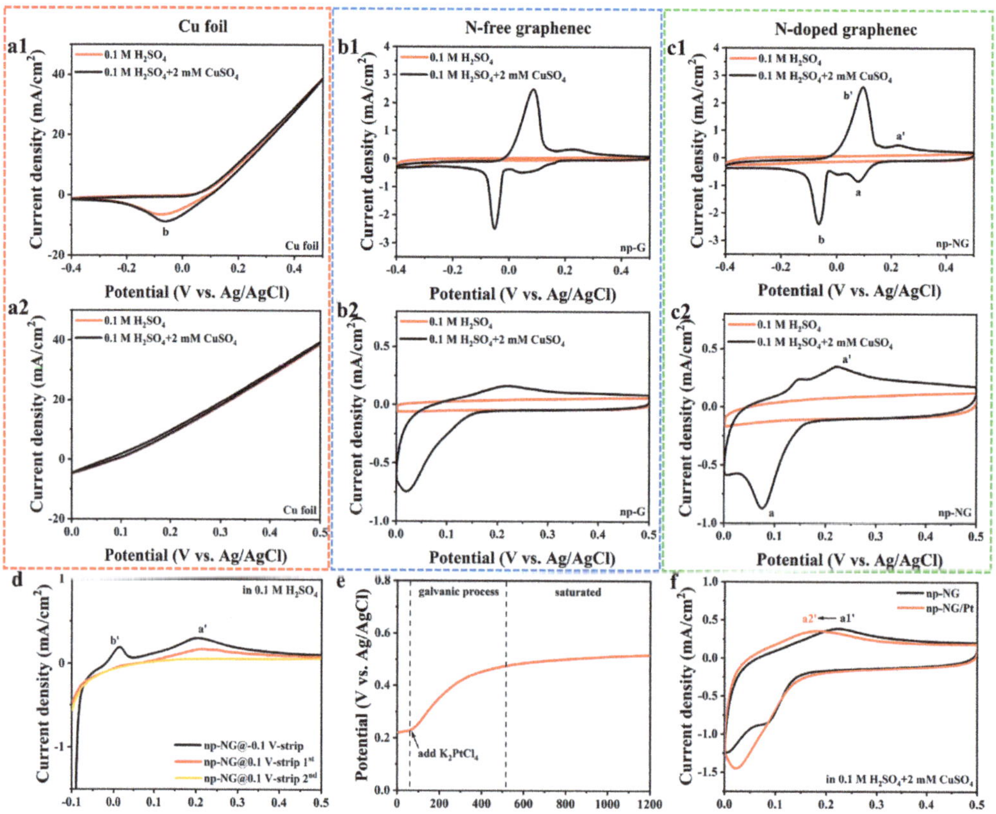

Figure 2. Electrochemical deposition of Cu on different samples. CV curves of Cu redox on (**a1**,**a2**) Cu foil, (**b1**,**b2**) np-G, and (**c1**,**c2**) np-NG samples over different potential windows. (**d**) LSV stripping curves of Cu deposited at different potentials. (**e**) Monitoring of the open circuit potential of the galvanic process. (**f**) CV curves of Cu redox on np-NG before and after loading Pt SAs.

The constant–potential deposition curves also indicate the difference between the two samples (Figure S3). The np-NG curve exhibits a clear UPD region, and the magnitude current gradually increases from −0.6 to 0 mA/cm^2 as UPD proceeds, indicating that the Cu single atoms gradually occupy the surface as specific sites [30]. However, the np-G curve does not display a UPD region and the current remains at 0 mA/cm^2 during the reaction, indicating that the number of sites that may be deposited on its surface is considerably less than that of np-NG. The voltage at which Cu bulk deposition occurs on np-NG was also chosen for potentiostatic deposition. The bulk deposition current is significantly larger than the UPD current, which coincides with the polarization curve. Stripping linear sweep voltammetry was performed after electrodeposition by transferring the samples to a Cu^{2+}-free 0.1 M H_2SO_4 electrolyte to analyze the form of Cu deposition on np-NG. Double peaks of both bulk Cu and the Cu SA appeared in the bulk-deposited samples, while only a single peak of the Cu SA appeared in the UPD sample in the first run of stripping. It is shown that the Cu SA can be selectively obtained by controlling the deposition potential. In the second run of the Cu SA stripping curve, the Cu SA peak in the UPD sample had completely disappeared, indicating the single-layer structure. The change in the open circuit potential of np-NG/Cu before and after the addition of 5 mM K_2PtCl_4 in a 0.05 M H_2SO_4 solution was monitored. The addition of K_2PtCl_4 at 60 s, followed by a slow rise in open circuit potential, indicated that the displacement of Cu occurred on the sample. The open circuit potential gradually saturated after about 500 s, indicating that Cu was almost completely replaced [17]. CV curve analysis of the redox reaction of Cu on the np-NG/Pt SA samples showed that the intensity of the oxidation peak of Cu was almost unchanged compared to the pristine np-NG, but the position was negatively shifted from a1′ (0.22 V) to a2′ (0.2 V). This indicates that the original site has been covered by Pt and the change in the work function of the new site caused the change in redox potential.

We investigated the chemical configurations and local coordination of different SA samples on the np-NG substrate using X-ray photoelectron (XPS) and X-ray absorption spectroscopy (XAS). Changes in N sites before and after UPD were evident by XPS. Based on the N 1s spectrum of pristine np-NG shown in Figure 3a, four different types of N are observed, namely pyridine-like (P-N, 398.6 eV, 62.4%), pyrrole-like (Py-N, 400.4 eV, 23.9%), graphitic (G-N, 401.6 eV, 12.8%), and oxidized (O-N, 403.3 eV, 0.9%) [39]. Remarkably, the percentage of P-N decreases significantly to 21.1% with the deposition of Cu SAs. A new metal–N peak (M-N) is observed at 399.9 eV, suggesting that the sites for the UPD of single atoms are mainly P-N sites, and the P-N sites are transformed into M-N sites after the generation of single atoms [40,41]. The N 1s spectrum of the sample after galvanic replacement are similar to that after the UPD of Cu, with a decrease in the P-N content and the appearance of the M-N peak, which suggests that interactions between the single atoms and the N sites still exist after replacement. When comparing the Cu 2p spectra before and after galvanic replacement (Figure 3b), a distinct Cu 2p peak is observed in the spectrum of np-NG/Cu. The $Cu^{\delta+}$ $2p_{3/2}$ peak is located at 933.08 eV between the peak positions for Cu(0) (932.60 eV) and Cu(II) (934.97 eV), indicating that $0 < \delta(Cu) < 2$, and electron transfer between Cu and N occurs. The Cu 2p signal almost disappears in the spectrum of the replaced sample, indicating that the Cu has been replaced. The $4f_{7/2}$ peak of Pt(0) is located at 71.2 eV and thus, the Pt 4f XP spectra shown in Figure 3c reveal that the Pt SAs occur in their oxidized states. Additionally, the $Pt^{\delta+}$ $4f_{7/2}$ peak in the spectrum of np-NG/Pt is located at 72.65 eV, which also indicates that $0 < \delta(Pt) < 2$ [42]. Hence, the strong electrostatic effects between the single atoms generated via galvanic replacement and the N defects on the support are still retained.

Based on X-ray absorption near-edge structure (XANES) spectroscopy, Pt L_3-edge analysis (Figure 3d) shows that the white line intensities of np-NG/Pt are higher than those of Pt foils but lower than those of PtO_2, suggesting that the Pt SAs display more unoccupied 5d orbitals compared to Pt(0) but do not reach valences of +4 [43]. The first-derivative XANES spectra shown in Figure S3 reveal that the valence states of both types of Pt SAs are between those of the reference Pt foil and PtO_2. Based on the fitting of the specific oxidation states using the reference Pt foils and PtO_2 (Figure 3e), the valence state of Pt in np-NG/Pt is +1.48, which is consistent with the XPS results. The local atom structure was further determined using extended X-ray absorption fine structure (EXAFS) analysis (Figure 1f). No Pt–Pt bonds are observed in np-NG/Pt, and a peak at 1.96 Å is observed instead. This peak may be attributed to Pt-N and Pt-C.

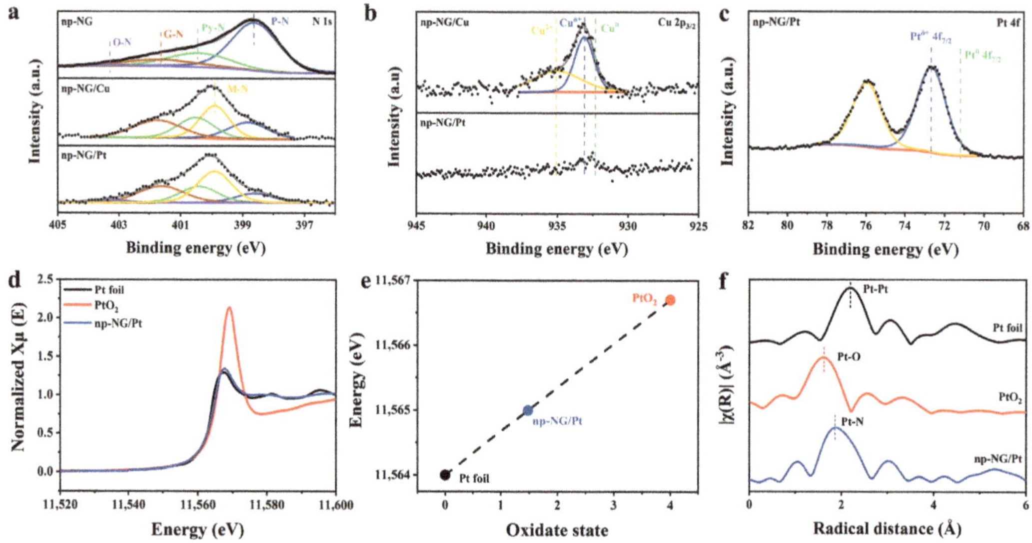

Figure 3. Structural characterization of different samples. XPS spectra of (**a**) N 1s, (**b**) Cu 2p, and (**c**) Pt 4f of different samples. (**d**) XANES and (**e**) correlation between the oxidation state and energy position. (**f**) Fourier transform EXAFS spectra of different Pt samples.

We have evaluated the HER performance of galvanically replaced np-NG/Pt SACs using a three-electrode system in the 0.5 M H_2SO_4 electrolyte. The Pt atom-loaded samples exhibit superior electrocatalytic HER activities compared to that of pristine np-NG. The different Pt catalysts exhibit different acidic HER activities in the order of commercial Pt/C (39 mV) > np-NG/Pt (47 mV) > np-G/Pt (211 mV) at a current density of 10 mA/cm^2 (Figure 4a). The np-NG (508 mV) as substrates do not exhibit HER activity. The performance of np-NG/Pt slightly decreased after 1000 cycle tests (53 mV@10 mA/cm^2), which may be due to the detachment of some poorly bound Pt SAs (Figure S4). To investigate the mechanisms of different samples in the HER under acidic conditions, we evaluated the catalytic kinetics using Tafel and Nyquist plots (Figure 4b,c). In all samples, the Tafel slope of np-NG/Pt (47.2 mV/dec) is close to that of commercial Pt/C (37.0 mV/dec) [44] and better than that of pristine np-NG (343.4 mV/dec), indicating that it exhibits good HER kinetics. Additionally, np-NG/Pt displays small charge transfer resistance (R_{ct}) under acidic conditions (overpotential of 50 mV), and the order of R_{ct} measured at an overpotential of 50 mV is Pt/C (10.1 Ω) < np-NG/Pt (11.5 Ω) < np-NG (600.1 Ω), suggesting that electrons can be transferred rapidly across np-NG/Pt.

In contrast to commercial Pt/C electrodes, np-NG/Pt did not show significant characteristic redox peaks in the Pt-H adsorption/desorption region (Figure 4d). For bulk Pt materials, there is a strong association between the H_{upd} peak and the surface structure, which can be used as a fingerprint of the surface orientation [45]. This suggests that it is possible that firstly, the ultra-low Pt loading (2 wt.% by XPS) in the Pt SAC exceeds the detection limit of cyclic voltammetry and secondly, there are differences in the characteristic redox peaks of the Pt SAs with respect to the bulk material. We compared the TOF of the Pt SAC with that of commercial Pt/C. The number of active sites in commercial Pt/C was obtained based on the ECSA, whereas in the Pt SAC, it is approximated that each atom is an active site. The densities of active sites in Pt/C and np-NG/Pt were 2.02×10^{-8} mol/cm^2 and 1.05×10^{-8} mol/cm^2, respectively, indicating that the Pt atoms in np-NG/Pt were well utilized. And at a 100 mV overpotential, the TOF of np-NG/Pt is 25.1 s^{-1}, which is higher than that of commercial Pt/C at 19.7 s^{-1} (Figure 4e), suggesting that np-NG/Pt has better intrinsic activity. Meanwhile, np-NG/Pt exhibits excellent mass activity due to the low noble metal loading, with a mass activity of 24.7 A/mg_{Pt} at a 100 mV overpotential, which is 6.7 times higher than that of commercial Pt/C. Several experimental and simulation results have shown that by lowering the oxidation state of Pt, the adsorption/desorption of H on the catalyst surface can be improved, thus enhancing the acidic HER performance of Pt [46–49]. This is consistent with the results we obtained in XPS and XAFS, where platinum has a lower valence and therefore exhibits good HER activity.

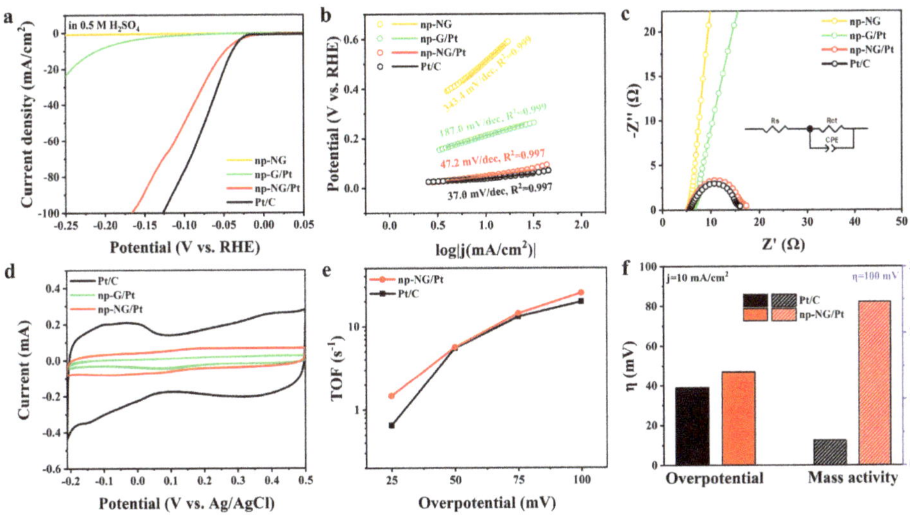

Figure 4. Electrochemical studies in 0.5 M H_2SO_4 electrolyte. (**a**) HER curves, (**b**) Tafel plots, (**c**) Nyquist plots, (**d**) ECSA curves, (**e**) TOF curves, and (**f**) overpotential (j = 10 mA/cm^2) and mass activities (η = 100 mV) of different catalysts.

4. Conclusions

In conclusion, we developed a method to prepare Cu SAs on N-doped graphene by UPD. The conditions are milder and more energy efficient than conventional methods of preparing SAs. Fine XPS and XAFS were used for the identification of SAs and the corresponding anchored specific sites. It was found that among all the N sites, pyridine-like N defects are the specific sites for the UPD of Cu SAs. Furthermore, by the galvanic process method, Pt replaces Cu to obtain np-NG/Pt SAC. The EMSIs between the specific sites and the SAs were able to effectively immobilize the SAs and regulate the oxidation state of the Pt SACs, which showed good HER performance under acidic conditions. Although the loading of Pt SAs was so small that it could not even be detected by H absorption/desorption, it exhibited a TOF of 25.1 s^{-1} and mass activity of 126.5 mA/mg_{Pt} at 100 mV. This method provides a new understanding and design ideas for the bottom-up preparation of atomic structures on N-doped graphene.

Supplementary Materials: The following supporting information can be downloaded at: https://www.mdpi.com/article/10.3390/ma17205082/s1, Figure S1: SEM images of np-Ni template (a) before and (b) after growing graphene and (c) np-NG after etching Ni template. (d) STEM and (e) TEM images of np-NG. (f) HRTEM image of the cross-section of np-NG; Figure S2: CV curves of different samples in 0.5 M H_2SO_4 containing 2 mM $CuSO_4$; Figure S3: Chronoamperometry curves of the deposition of Cu at different potentials; Figure S4: The corresponding first-derivative XANES spectra of different Pt samples; Figure S5: Stability test of np-NG/Pt by potential cycling before and after 1000 cycles.

Author Contributions: Conceptualization, H.W. and P.L.; validation, H.W. and Q.Z.; formal analysis, S.C.; resources, H.D., Y.W. and P.L.; data curation, H.W.; writing—original draft preparation, H.W.; writing—review and editing, P.L.; funding acquisition, H.D., Y.W. and P.L. All authors have read and agreed to the published version of the manuscript.

Funding: This work was supported by the National Natural Science Foundation of China (grant no. 52173224, 52130105) and the Natural Science Foundation of Shanghai (grant no. 23ZR1431500). This work was also supported by the User Experiment Assist System of SSRF. This work was also supported by Shanghai Jiao Tong University—JA Solar New Energy Materials Joint Research Center.

Institutional Review Board Statement: Not applicable.

Informed Consent Statement: Not applicable.

Data Availability Statement: The original contributions presented in the study are included in the article/Supplementary Materials, further inquiries can be directed to the corresponding author.

Conflicts of Interest: Authors Hao Du, Yanyue Wang and Pan Liu were employed by the company JA Solar Technology Co., Ltd. The remaining authors declare that the research was conducted in the absence of any commercial or financial relationships that could be construed as a potential conflict of interest.

References

1. Zhou, P.; Niu, P.; Liu, J.; Zhang, N.; Bai, H.; Chen, M.; Feng, J.; Liu, D.; Wang, L.; Chen, S.; et al. Anodized steel: The most promising bifunctional electrocatalyst for alkaline water electrolysis in industry. *Adv. Funct. Mater.* **2022**, *32*, 2202068. [CrossRef]
2. Chatenet, M.; Pollet, B.G.; Dekel, D.R.; Dionigi, F.; Deseure, J.; Millet, P.; Braatz, R.D.; Bazant, M.Z.; Eikerling, M.; Staffell, I.; et al. Water electrolysis: From textbook knowledge to the latest scientific strategies and industrial developments. *Chem. Soc. Rev.* **2022**, *51*, 4583–4762. [PubMed]
3. Sun, H.N.; Xu, X.M.; Kim, H.S.; Jung, W.C.; Zhou, W.; Shao, Z.P. Electrochemical water splitting: Bridging the gaps between fundamental research and industrial applications. *Energy Environ. Mater.* **2023**, *6*, e12441. [CrossRef]
4. Morales-Guio, C.G.; Stern, L.A.; Hu, X.L. Nanostructured hydrotreating catalysts for electrochemical hydrogen evolution. *Chem. Soc. Rev.* **2014**, *43*, 6555–6569. [CrossRef]
5. Zhang, J.; Wang, T.; Liu, P.; Liao, Z.; Liu, S.; Zhuang, X.; Chen, M.; Zschech, E.; Feng, X. Efficient hydrogen production on $MoNi_4$ electrocatalysts with fast water dissociation kinetics. *Nat. Commun.* **2017**, *8*, 15437. [CrossRef]
6. Dresselhaus, M.S.; Thomas, I. Alternative energy technologies. *Nature* **2001**, *414*, 332–337. [CrossRef]
7. Nørskov, J.K.; Bligaard, T.; Logadottir, A.; Kitchin, J.; Chen, J.G.; Pandelov, S.; Stimming, U. Trends in the exchange current for hydrogen evolution. *J. Electrochem. Soc.* **2005**, *152*, J23. [CrossRef]
8. Hansen, J.N.; Prats, H.; Toudahl, K.K.; Mørch Secher, N.; Chan, K.; Kibsgaard, J.; Chorkendorff, I. Is there anything better than Pt for HER? *ACS Energy Lett.* **2021**, *6*, 1175–1180. [CrossRef]
9. Hong, Y.-R.; Dutta, S.; Jang, S.W.; Ngome Okello, O.F.; Im, H.; Choi, S.-Y.; Han, J.W.; Lee, I.S. Crystal facet-manipulated 2D Pt danodendrites to achieve an intimate heterointerface for hydrogen evolution reactions. *J. Am. Chem. Soc.* **2022**, *144*, 9033–9043. [CrossRef]
10. Ouyang, Y.R.; Cao, H.J.; Wu, H.J.; Wu, D.B.; Wang, F.Q.; Fan, X.J.; Yuan, W.Y.; He, M.X.; Zhang, L.Y.; Li, C.M. Tuning Pt-skinned PtAg nanotubes in nanoscales to efficiently modify electronic structure for boosting performance of methanol electrooxidation. *Appl. Catal. B-Environ.* **2020**, *265*, 118606. [CrossRef]
11. Kitchin, J.R.; Nørskov, J.K.; Barteau, M.A.; Chen, J.G. Modification of the surface electronic and chemical properties of Pt(111) by subsurface 3d transition metals. *J. Chem. Phys.* **2004**, *120*, 10240–10246. [CrossRef] [PubMed]
12. Wang, X.S.; Zheng, Y.; Sheng, W.C.; Xu, Z.J.; Jaroniec, M.; Qiao, S.Z. Strategies for design of electrocatalysts for hydrogen evolution under alkaline conditions. *Mater. Today* **2020**, *36*, 125–138. [CrossRef]
13. Singh, S.B.; De, M. Thermally exfoliated graphene oxide for hydrogen storage. *Mater. Chem. Phys.* **2020**, *239*, 122102. [CrossRef]
14. Singh, S.B.; De, M. Scope of doped mesoporous (<10 nm) surfactant-modified alumina templated carbons for hydrogen storage applications. *Int. J. Energy Res.* **2019**, *43*, 4264–4280.
15. Campbell, C.T. Electronic perturbations. *Nat. Chem.* **2012**, *4*, 597–598. [CrossRef]
16. Hu, P.P.; Huang, Z.W.; Amghouz, Z.; Makkee, M.; Xu, F.; Kapteijn, F.; Dikhtiarenko, A.; Chen, Y.; Gu, X.; Tang, X.F. Electronic metal-support interactions in single-atom catalysts. *Angew. Chem. Int. Edit.* **2014**, *53*, 3418–3421. [CrossRef]
17. Bezerra, C.; Mamedov, D.; Alonso-Vante, N. The effect of platinum single atoms in titania for photocatalytic applications. *Johnson Matthey Tech.* **2024**, *68*, 201. [CrossRef]
18. Wang, A.Q.; Li, J.; Zhang, T. Heterogeneous single-atom catalysis. *Nat. Rev. Chem.* **2018**, *2*, 65–81. [CrossRef]
19. Cheng, N.C.; Zhang, L.; Doyle-Davis, K.; Sun, X.L. Single-atom catalysts: From design to application. *Electrochem. Energy Rev.* **2019**, *2*, 539–573. [CrossRef]
20. Zhu, C.Z.; Fu, S.F.; Shi, Q.R.; Du, D.; Lin, Y.H. Single-atom electrocatalysts. *Angew. Chem. Int. Edit.* **2017**, *56*, 13944–13960. [CrossRef]
21. Li, L.; Wang, B.; Zhang, G.W.; Yang, G.; Yang, T.; Yang, S.; Yang, S.C. Electrochemically modifying the electronic structure of IrO_2 nanoparticles for overall electrochemical water splitting with extensive adaptability. *Adv. Energy Mater.* **2020**, *10*, 2001600. [CrossRef]

22. Shi, Y.; Ma, Z.R.; Xiao, Y.Y.; Yin, Y.C.; Huang, W.M.; Huang, Z.C.; Zheng, Y.Z.; Mu, F.Y.; Huang, R.; Shi, G.Y.; et al. Electronic metal-support interaction modulates single-atom platinum catalysis for hydrogen evolution reaction. *Nat. Commun.* **2021**, *12*, 3021. [CrossRef] [PubMed]
23. Li, J.J.; Guan, Q.Q.; Wu, H.; Liu, W.; Lin, Y.; Sun, Z.H.; Ye, X.X.; Zheng, X.S.; Pan, H.B.; Zhu, J.F.; et al. Highly active and stable metal single-atom catalysts achieved by strong electronic metal–support interactions. *J. Am. Chem. Soc.* **2019**, *141*, 14515–14519. [CrossRef] [PubMed]
24. Shah, K.; Dai, R.; Mateen, M.; Hassan, Z.; Zhuang, Z.; Liu, C.; Israr, M.; Cheong, W.C.; Hu, B.; Tu, R.; et al. Cobalt single atom incorporated in ruthenium oxide sphere: A robust bifunctional electrocatalyst for HER and OER. *Angew. Chem. Int. Edit.* **2021**, *134*, e202114951. [CrossRef]
25. Jin, H.; Sultan, S.; Ha, M.; Tiwari, J.N.; Kim, M.G.; Kim, K.S. Simple and Scalable Mechanochemical Synthesis of Noble Metal Catalysts with Single Atoms toward Highly Efficient Hydrogen Evolution. *Adv. Funct. Mater.* **2020**, *30*, 2000531. [CrossRef]
26. Qu, Y.; Chen, B.; Li, Z.; Duan, X.; Wang, L.; Lin, Y.; Yuan, T.; Zhou, F.; Hu, Y.; Yang, Z.; et al. Thermal emitting strategy to synthesize atomically dispersed Pt metal sites from bulk Pt metal. *J. Am. Chem. Soc.* **2019**, *141*, 4505–4509. [CrossRef]
27. Li, J.; Yin, H.-M.; Li, X.-B.; Okunishi, E.; Shen, Y.-L.; He, J.; Tang, Z.-K.; Wang, W.-X.; Yücelen, E.; Li, C.; et al. Surface evolution of a Pt–Pd–Au electrocatalyst for stable oxygen reduction. *Nat. Energy* **2017**, *2*, 1–9. [CrossRef]
28. Shi, Y.; Huang, W.M.; Li, J.; Zhou, Y.; Li, Z.Q.; Yin, Y.C.; Xia, X.H. Site-specific electrodeposition enables self-terminating growth of atomically dispersed metal catalysts. *Nat. Commun.* **2020**, *11*, 4558. [CrossRef]
29. Meese, A.F.; Napier, C.; Kim, D.J.; Rigby, K.; Hedtke, T.; Leshchev, D.; Stavitski, E.; Parent, L.R.; Kim, J.H. Underpotential Deposition of 3D Transition Metals: Versatile Electrosynthesis of Single-Atom Catalysts on Oxidized Carbon Supports. *Adv. Mater.* **2024**, *36*, 2311341. [CrossRef]
30. Xu, J.; Li, R.; Xu, C.-Q.; Zeng, R.; Jiang, Z.; Mei, B.; Li, J.; Meng, D.; Chen, J. Underpotential-deposition synthesis and in-line electrochemical analysis of single-atom copper electrocatalysts. *Appl. Catal. B-Environ.* **2021**, *289*, 120028. [CrossRef]
31. Cheng, Y.T.; Wu, H.F.; Han, J.H.; Zhong, S.Y.; Huang, S.H.; Chu, S.F.; Song, S.X.; Reddy, K.M.; Wang, X.D.; Wu, S.Y.; et al. Atomic Ni and Cu co-anchored 3D nanoporous graphene as an efficient oxygen reduction electrocatalyst for zinc-air batteries. *Nanoscale* **2021**, *13*, 10862–10870. [CrossRef] [PubMed]
32. Ito, Y.; Qiu, H.J.; Fujita, T.; Tanabe, Y.; Tanigaki, K.; Chen, M.W. *Bicontinuous nanoporous* N-doped graphene for the oxygen reduction reaction. *Adv. Mater.* **2014**, *26*, 4145–4150. [CrossRef] [PubMed]
33. Qiu, Y.; Lopez-Ruiz, J.A.; Zhu, G.; Engelhard, M.H.; Gutiérrez, O.Y.; Holladay, J.D. Electrocatalytic decarboxylation of carboxylic acids over RuO2 and Pt nanoparticles. *Appl. Catal. B-Environ.* **2022**, *305*, 121060. [CrossRef]
34. Kibsgaard, J.; Jaramillo, T.F. Molybdenum phosphosulfide: An active, acid-stable, earth-abundant catalyst for the hydrogen evolution reaction. *Angew. Chem. Int. Edit.* **2014**, *53*, 14433–14437. [CrossRef]
35. Estudillo-Wong, L.A.; Alonso-Vante, N. Factors Intervening in Oxide and Oxide-Composite Supports on Nanocatalysts in the Energy Conversion. *Heterog. Nanocatal. Energy Environ. Sustain.* **2022**, *1*, 1–60.
36. Hussain, S.; Kongi, N.; Erikson, H.; Rähn, M.; Merisalu, M.; Matisen, L.; Paiste, P.; Aruväli, J.; Sammelselg, V.; Estudillo-Wong, L.A.; et al. Platinum nanoparticles photo-deposited on SnO_2-C composites: An active and durable electrocatalyst for the oxygen reduction reaction. *Electrochim. Acta* **2019**, *316*, 162–172. [CrossRef]
37. Zhang, Z.; Wang, P. Highly stable copper oxide composite as an effective photocathode for water splitting via a facile electro chemical synthesis strategy. *J. Mater. Chem.* **2012**, *22*, 2456–2464. [CrossRef]
38. Xu, J.; Li, R.; Zeng, R.; Li, Y.; Pan, Q.; Zhong, H.; Chen, J. Underpotential deposition of copper clusters on sulfur and nitrogen Co-doped graphite foam for the oxygen reduction reaction. *ChemElectroChem* **2019**, *6*, 5682–5687. [CrossRef]
39. Liu, J.; Jiao, M.G.; Lu, L.L.; Barkholtz, H.M.; Li, Y.P.; Wang, Y.; Jiang, L.H.; Wu, Z.J.; Liu, D.J.; Zhuang, L.; et al. High performance platinum single atom electrocatalyst for oxygen reduction reaction. *Nat. Commun.* **2017**, *8*, 15938. [CrossRef]
40. Wang, Y.; Yu, B.; Liu, K.; Yang, X.; Liu, M.; Chan, T.-S.; Qiu, X.; Li, J.; Li, W. Co single-atoms on ultrathin N-doped porous carbon via a biomass complexation strategy for high performance metal-air batteries. *J. Mater. Chem. A* **2020**, *8*, 2131–2139. [CrossRef]
41. Chen, G.; An, Y.; Liu, S.; Sun, F.; Qi, H.; Wu, H.; He, Y.; Liu, P.; Shi, R.; Zhang, J.; et al. Highly accessible and dense surface single metal FeN_4 active sites for promoting the oxygen reduction reaction. *Energy Environ. Sci.* **2022**, *15*, 2619–2628. [CrossRef]
42. Wagner, C. Chemical shifts of auger lines, and the auger parameter. *Faraday Discuss. Chem. Soc.* **1975**, *60*, 291–300. [CrossRef]
43. Zhang, Z.Q.; Liu, J.P.; Wang, J.; Wang, Q.; Wang, Y.H.; Wang, K.; Wang, Z.; Gu, M.; Tang, Z.H.; Lim, J.W.; et al. Single-atom catalyst for high-performance methanol oxidation. *Nat. Commun.* **2021**, *12*, 5235. [CrossRef] [PubMed]
44. Heijden, O.V.D.; Park, S.; Vos, R.E.; Eggebeen, J.J.J.; Koper, M.T.M. Tafel slope plot as a tool to analyze electrocatalytic reactions. *ACS Energy Lett.* **2024**, *9*, 1871–1879. [CrossRef]
45. Estudillo-Wong, L.A.; Ramos-Sanchez, G.; Calvillo, L.; Granozzi, G.; Alonso-Vante, N. Support interaction effect of platinum nanoparticles on non-, Y-, Ce-doped anatase and its implication on the ORR in acid and alkaline media. *ChemElectroChem* **2017**, *4*, 3264–3275. [CrossRef]
46. Yang, L.; Grzeschik, R.; Jiang, P.; Yu, L.F.; Hu, C.J.; Du, A.X.; Schlücker, S.; Xie, W. Tuning the electronic properties of platinum in hybrid-nanoparticle assemblies for use in hydrogen evolution reaction. *Angew. Chem. Int. Edit.* **2023**, *62*, e202301065. [CrossRef]
47. Yan, Q.Q.; Wu, D.X.; Chu, S.Q.; Chen, Z.Q.; Lin, Y.; Chen, M.X.; Zhang, J.; Wu, X.J.; Liang, H.W. Reversing the charge transfer between platinum and sulfur-doped carbon support for electrocatalytic hydrogen evolution. *Nat. Commun.* **2019**, *10*, 4977. [CrossRef]

48. Fan, X.L.; Liu, C.; Gao, B.X.; Li, H.; Zhang, Y.H.; Zhang, H.B.; Gao, Q.S.; Cao, X.M.; Tang, Y. Electronic structure engineering of Pt species over Pt/WO_3 toward highly efficient electrocatalytic hydrogen evolution. *Small* **2023**, *19*, 2301178. [CrossRef]
49. Ma, T.; Cao, H.; Li, S.; Cao, S.J.; Zhao, Z.Y.; Wu, Z.H.; Yan, R.; Yang, C.D.; Wang, Y.; van Aken, P.A.; et al. Crystalline Lattice-Confined Atomic Pt in Metal Carbides to Match Electronic Structures and Hydrogen Evolution Behaviors of Platinum. *Adv. Mater.* **2022**, *34*, 2206368. [CrossRef]

materials

MDPI

Article

Graphene Supported NiFe-LDH and PbO_2 Catalysts Prepared by Plasma Process for Oxygen Evolution Reaction

Tingting Yang [1], Zheng Zhang [1], Fei Tan [1], Huayu Liu [1], Xingyu Li [1], Hongqi Wang [2] and Qing Yang [3,*]

1 State Grid Chongqing Electric Power Company Material Branch, Chongqing 401120, China
2 Chongqing Jie Chuang Electric Power Technology Company Ltd., Chongqing 400031, China
3 State Key Laboratory of Power Transmission Equipment Technology, School of Electrical Engineering, Chongqing University, Chongqing 400044, China
* Correspondence: yangqing@cqu.edu.cn

Abstract: The development of efficient catalysts for water electrolysis is crucial for advancing the low-carbon transition and addressing the energy crisis. This work involves the fabrication of graphene-based catalysts for the oxygen evolution reaction (OER) by integrating NiFe-LDH and PbO_2 onto graphene using plasma treatment. The plasma process takes only 30 min. Graphene's two-dimensional structure increases the available reaction surface area and improves surface electron transport. Plasma treatment further improves catalyst performance by facilitating nanoparticle attachment and creating carbon defects and sulfur vacancies. Density functional theory (DFT) calculations at the PBE provide valuable insights into the role of vacancies in enhancing catalyst performance for OER. The catalyst's conductivity and electronic structure are greatly impacted by vacancies. While modifications to the electronic structure increase the kinetics of charge transfer, the vacancy structure can produce more active sites and improve the adsorption and reactivity of OER intermediates. This optimization of intermediate adsorption and electronic properties leads to increased overall OER activity. The catalyst NiFe-PbO_2/S/rGO-45, synthesized through plasma treatment, demonstrated an overpotential of 230 mV at 50 $mA{\cdot}cm^{-2}$ and a Tafel slope of 44.26 mV dec^{-1}, exhibiting rapid reaction kinetics and surpassing the OER activity of commercial IrO_2. With its excellent performance, the prepared catalyst has broad prospects in commercial applications such as water electrolysis and air batteries.

Keywords: plasma applications; graphene; vacancies; electrocatalysis; OER

1. Introduction

To meet the demands of energy structure adjustment and ensure the safe and stable operation of power grids, the integration of renewable energy sources is becoming increasingly imperative. A potential energy source, hydrogen is capable of being created from a variety of non-renewable and renewable resources. Hydrogen is distinguished for its high calorific value and environmentally benign combustion properties, as it only produces water as a byproduct when used in fuel cells or combustion engines. These characteristics make hydrogen an attractive option [1–3]. Water electrolysis, particularly when powered by renewable energy, is a key technology for achieving carbon peak and carbon neutrality goals. Nonetheless, the OER occurring at the anode during water electrolysis poses a considerable challenge. This process involves a complex four-electron transfer, which is kinetically slow, requires a high overpotential, and results in low reaction efficiency—factors that hinder the widespread adoption of water electrolysis technology [4,5]. To tackle these challenges, effective OER catalysts that could enhance performance and reduce the reaction energy barrier require being developed.

Currently, most OER catalysts rely on precious metals or their compounds. Despite the high catalytic activity, high cost and limited availability hinder large-scale applications [6,7]. In contrast, transition metals are garnering significant interest because of their plentiful reserves, affordability, and effective catalytic properties. Nickel-iron bimetallic catalysts, in particular, have gained widespread interest for their high catalytic activity and relatively lower cost [8,9]. The combined influence of nickel and iron lowers the energy barrier of OER, thereby enhancing catalytic efficiency. However, NiFe bimetallic catalysts still face several practical challenges, such as poor stability of active sites, insufficient electrical conductivity, and reduced catalytic activity caused by the formation of surface oxides. To overcome these issues, recent research has suggested that incorporating PbO_2 and graphene into the catalyst could significantly improve its performance. PbO_2 is utilized because of its high conductivity, which compensates for the poor conductivity of transition metals. At the same time, since lead dioxide can be obtained from waste lead-acid batteries, the use of lead dioxide can promote the recycling of lead-acid batteries [10]. Graphene is known for its excellent redox properties and high electrical conductivity. The combination of PbO_2 and graphene helps enhance charge transfer efficiency, stabilize the catalyst surface, and alleviate the decrease of active sites, consequently improving the durability and catalytic efficacy [11,12]. Furthermore, the development of vacancy structures in graphene and the associated compounds during synthesis has the potential to increase the reaction surface area, thereby improving catalytic efficiency. For example, Rocha et al. fabricated a multi-vacancy, doped low-concentration nickel-reduced graphene oxide 3D aerogel through freeze-casting, achieving an overpotential of 480 mV [13]. Fei et al. created a porous core-shell nanoreactor that encapsulates NiFe within graphene layers. This design exhibited remarkable stability, operating for 2000 h under alkaline seawater conditions [14]. These studies highlight the potential of PbO_2 and graphene, along with vacancy structures, to address the limitations of transition metals in OER catalysts. However, the underlying mechanisms of these improvements still require further investigation.

Herein, this work examines the impact of vacancy structures on the performance of graphene-supported transition metals and PbO_2, employing RF plasma treatment to integrate NiFe-LDH and PbO_2 onto graphene. The analysis focuses on the morphological structure and electrochemical performance, aiming to uncover their surface catalytic activity and offer insights into the underlying microscopic reaction mechanisms.

2. Materials and Methods

2.1. Materials

The materials utilized include commercial graphene oxide powder (GO, purity 99.5%, thickness 6–8 nm, width 25 μm) sourced from $Ni(NO_3)_2{\cdot}6H_2O$, $FeSO_4{\cdot}7H_2O$, thiourea, and PbO_2, all of analytical grade, also obtained from Macklin; and anhydrous ethanol, which is of analytical grade, acquired from Chengdu Kelong.

2.2. Preparation of NiFe-PbO_2/S/rGO-X

A precursor solution was prepared by adding the following materials to 30 mL of deionized water: 30 mg of graphene oxide, 145.395 mg of nickel nitrate hexahydrate, 139.025 mg of iron (II) sulfate heptahydrate, 111.6 mg of lead dioxide, and 228.36 mg of thiourea. To ensure complete dissolution, the solution was sonicated for half an hour. To get the precursor powder sample, the solution was then freeze-dried for 24 h at −75 °C.

An RF plasma-assisted treatment experimental platform was established, as shown in Figure 1. The system consists of a radio frequency (RF) power generator (Beijing Jinmao

Technology Co., Ltd., Beijing, China), a matching network, and a vacuum system, which work together to generate plasma for the treatment process. A copper plate wrapped around a quartz tube serves as the electrode, which is connected to the RF power supply. Argon gas (Ar, purity 99.999%) is introduced into the reaction chamber to generate the plasma. The treatment process begins by placing the precursor inside the chamber, followed by the introduction of argon gas under vacuum conditions at a flow rate of 50 sccm. The plasma is then produced utilizing an RF power supply functioning at a frequency of 13.56 MHz and a power output of 80 W. The precursor undergoes plasma treatment for varying durations, which allows the control of vacancy structure development. NiFe-PbO_2/S/rGO-X catalyst samples are obtained, where "X" refers to the treatment time of plasma (X = 15, 30, 45, 60, min), with each sample corresponding to a specific degree of vacancy structure formation (Figure 2).

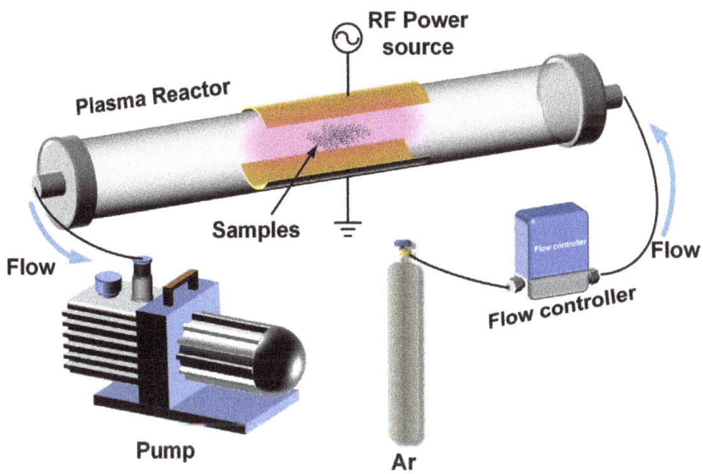

Figure 1. RF plasma experimental platform.

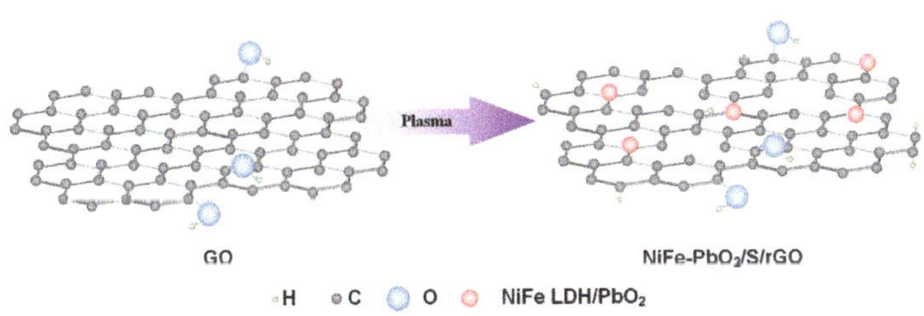

Figure 2. Preparation process of NiFe-PbO_2/S/rGO-X catalyst.

2.3. Material Characterization

The catalyst was characterized using a variety of analytical techniques to study its structure and composition. X-ray diffraction (XRD) research was conducted utilizing a D8 Advance (Bruker, Bremen, Germany) at a scanning velocity of 0.2°/s to investigate the crystalline structure. Raman spectra were acquired using a 532 nm laser Raman microscope equipment (Renishaw, London, UK) to examine the crystal phase. Scanning electron microscopy (SEM) pictures were obtained with a S-4800 (Hitachi, Tokyo, Japan) to examine the surface morphology of the catalyst. X-ray photoelectron spectroscopy (XPS) was performed using an ESC-Lab-250Xi (Thermo Fisher Scientific Inc., Waltham, MA, USA) with monochromatic Al-Kα radiation to analyze the chemical states of the elements. Electron paramagnetic resonance (EPR) tests utilizing the A30010/12 system

(Bruker, Bremen, Germany) were conducted to investigate the structure of the detector. The elemental concentrations were measured using inductively coupled plasma mass spectrometry (ICP-MS, ICP OES 7200, Thermo Fisher Scientific Inc., Waltham, MA, USA). The emission spectra of the materials were acquired utilizing an MX2500+ spectrometer (Ocean Optics, Orlando, FL, USA), covering a wavelength range of 200 nm to 1100 nm.

2.4. Preparation of the Ink of Working Electrode

Prepare a solution by dissolving 10 mg of the catalyst sample in a combination of 400 μL anhydrous ethanol and 40 μL 0.5% Nafion solution. Subject the mixture to sonication for a duration of 30 min, then permit it to stand for subsequent utilization. To facilitate comparison, dissolve 10 mg of IrO_2 (sourced from Shanghai Macklin, analytical grade) in the identical solvent mixture—400 μL anhydrous ethanol and 40 μL 0.5% Nafion solution. Sonicate for 30 min and set aside for later use.

2.5. Electrochemical Characterization

Electrochemical performance evaluation was conducted with a three-electrode configuration on a CHI 760 (Shanghai Chenhua Instrument Co. Ltd, Shanghai, China). The working electrode was fabricated by depositing 100 μL catalyst slurry onto a 1×1 cm^2 nickel foam substrate. A platinum electrode functioned as the counter electrode, whereas a silver/silver chloride electrode was employed as the reference electrode. All experiments were performed in an O_2-saturated 1 M KOH. The calibration equation employed for potential measurements was $E_{RHE} = E_{Ag/AgCl} + 0.0591 \times pH + 0.197$. Cyclic voltammetry (CV) measurements were conducted at a scan rate of 10 mV s^{-1} to evaluate the electrochemical properties of the catalyst. Electrochemical impedance spectroscopy (EIS) was conducted over a frequency range of 10^{-2} Hz to 10^5 Hz to examine the correlation between impedance and OER at different applied potentials. The current density was assessed for a duration of 50 h by amperometric i–t curves at a constant voltage of 1.52 V to evaluate long-term stability. The catalyst's OER activity was further investigated by linear sweep voltammetry (LSV) at a scan rate of 5 mV s^{-1}, within a potential range of 1–2 V vs. RHE. The double-layer capacitance (C_{dl}) method was employed to investigate the catalyst's electrochemical active surface area (ECSA) and improve catalytic performance. CV measurements were performed at many scan speeds inside an electrochemical window devoid of redox processes, facilitating the determination of C_{dl}.

2.6. Computational Details

Density functional theory (DFT) was performed to determine the energy and structure of each OER intermediate in two heterostructure models: NiFe-PbO_2/S/V_G and V_G and Graphene. In these models, plasma deposition is tackled by the DFT-backed synthetic growth concept; NiFe and PbO_2 were supported on a (4×4) supercell of graphene with vacancy defects. An optimized model of graphene, an optimized model of vacancy-defect graphene, was established for comparison. The layered two-dimensional structure of graphene was established based on the reported research [15,16]. The substrates were formed by NiFe (003) and PbO_2 (111) crystal planes. The computations were performed utilizing CASTEP with the Perdew-Burke-Ernzerhof (PBE) exchange-correlation functional under the generalized gradient approximation (GGA). The U_{eff} values for nickel and iron were established at 6.6 eV and 3.5 eV, respectively. Additionally, we used the Nose-Hoover thermostats at 300 K and 600 K, respectively, to perform ab initio molecular dynamics (AIMD) simulations in the NVT canonical ensembles. For the AIMD simulations, the time step was 2.0 fs and the overall time was 10 ps.

3. Results and Discussion

XRD and Raman were employed to examine the structural properties of NiFe-PbO_2/S/rGO-X. As shown in Figure 3a, XRD analysis of NiFe-PbO_2/S/rGO-45 reveals several diffraction peaks corresponding to NiFe LDH (JCPDS No. 40-0215) at 11.4°, 22.9°, 34.4°, and 59.8°, which correspond to the (003), (006), (012), and (110) crystal planes [17]. Additionally, peaks at 28.5°, 30°, 32.7°, 36°, 50.6°, and 55.8° correspond to the (111), (020), (002), (200), (221), and (113) crystal planes of PbO_2 (JCPDS No. 37-0517), confirming the loading of NiFe-LDH and PbO_2 onto the graphene surface.

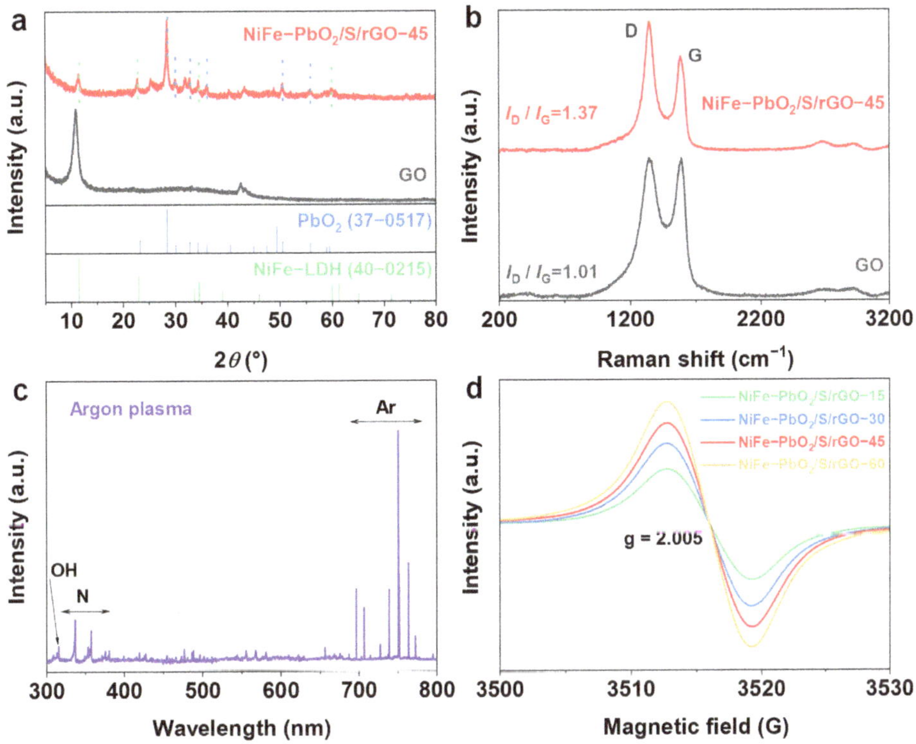

Figure 3. (**a**) XRD patterns of NiFe-PbO_2/S/rGO-45, (**b**) Raman spectra of NiFe-PbO_2/S/rGO-45 and GO, (**c**) Optical emission spectrum of Argon plasma, (**d**) EPR spectrum of NiFe-PbO_2/S/rGO-X.

Raman spectroscopy further examined the defective structure of the graphene. The D peak at 1350 cm^{-1}, which indicates the presence of defects, and the G peak at 1580 cm^{-1}, which represents the graphitic structure, demonstrate that plasma treatment induces defects in the graphene [18,19]. The intensity ratio of the D peak to the G peak (I_D/I_G) for NiFe-PbO_2/S/rGO-45 is 1.37. This value markedly surpasses the ratio of 1.01 for graphene oxide (GO), suggesting that the Ar plasma treatment effectively etches the graphene, creating a large number of defects while preserving its carbon framework. These defects enhance ion adsorption, facilitate carrier migration, and provide sites for the accommodation of NiFe LDH and PbO_2 nanoparticles. Additionally, the plasma treatment modifies the graphene's electronic band structure, promoting hole-type carriers and increasing the band gap, which facilitates sulfur doping and the formation of catalytic active sites [20]. Furthermore, carbon vacancies in graphene-based materials can also enhance their catalytic activity towards the OER. Highly localized states close to the Fermi level are introduced by carbon vacancies, and these states can operate as active sites for the adsorption and activation of OER intermediates [21]. The highly localized states introduced by carbon vacancies facilitate the adsorption and activation of OER intermediates, reducing the overpotential required for the reaction, and act as anchoring sites for metal nanoparticles, promoting the dispersion and stability of the catalytically active phase, further enhancing OER activity [22].

OES is an efficient technique for analyzing the active species and their physicochemical characteristics (Figure 3c). It offers significant insights into the origin of active particles in the sample, facilitating the comprehension of the effects of plasma on the NiFe-PbO_2/S/rGO-X lattice [23]. Plasma treatment with argon dramatically modifies the properties of the lattice by introducing high-energy particles and reactive functional groups (Ar, -OH, -N). The species in question are capable of initiating redox reactions, which leads to the formation of sulfur vacancies as catalytic active sites, crucial for improving the efficacy of the catalyst [24].

The catalyst's surface is struck by the charged particles in the plasma, which promotes atomic sputtering and increases the number of vacancy defects. The high temperature of plasma promotes the formation of intermediate products such as H_2S and NH_3 from thiourea, while plasma-induced ionization and excitation of gas molecules facilitate S and N doping into the graphene structure. Furthermore, free electrons, ions, and atoms inside the plasma engage with the GO surface, adsorbing negatively charged oxygen functional groups that subsequently undergo reduction, hence augmenting the material's conductivity. The plasma etching effect creates nanopores on the rGO surface, providing low-resistance channels for O_2 and increasing the surface area for metal oxide nanoparticle support.

Consequently, plasma treatment interferes with the C-S and S-S bonds in S-doped rGO, resulting in the creation of vacancies that modify the electronic structure. This technique facilitates concurrent material loading and etching, producing a synergistic action that boosts the OER catalytic efficacy. Post-plasma treatment, the number of active sites on the catalyst surface escalates, augmenting the reactive area and enhancing catalytic activity [25,26].

EPR measurements were conducted to further validate the existence of vacancy structures in the NiFe-PbO_2/S/rGO-X composite (Figure 3d). An identifiable S vacancy signal was at g = 2.005 [27]. S vacancies cause unsaturated sites to develop, which operate as active sites for the adsorption and activation of OER intermediates because they have higher electron densities and better charge transfer capacities [28]. This leads to a reduction in the overpotential required for OER, enhancing the catalytic activity [29]. Moreover, the intensity of the EPR signal increased with plasma etching duration, signifying a progressive rise in the number of vacancies. This indicates that modifying the plasma treatment duration allows for effective control over vacancy defects, which in turn promotes better electron transfer and enhanced catalytic activity.

To investigate the effect of plasma treatment on the microstructure of NiFe-PbO_2/S/rGO-X, SEM was used for characterization. As shown in Figure 4a,b, plasma treatment exfoliates GO into reduced graphene oxide (rGO) nanosheets. This reduction is driven by free active radicals from the Ar plasma, which interact with the GO surface. GO consists of both sp^2 and sp^3 domains, with sp^2 representing the unoxidized regions and sp^3 representing the oxidized regions. The high-energy particles of plasma erode GO, cleaving carbon-oxygen bonds and eliminating oxygen atoms. Furthermore, free electrons, ionized argon molecules, and atoms from the plasma engage with the graphene oxide surface, resulting in the desorption of oxygen functional groups and the diminution of sp^3 domains. This results in the reduction of GO, facilitated by the plasma treatment, which plays a crucial role in removing oxygenated functional groups.

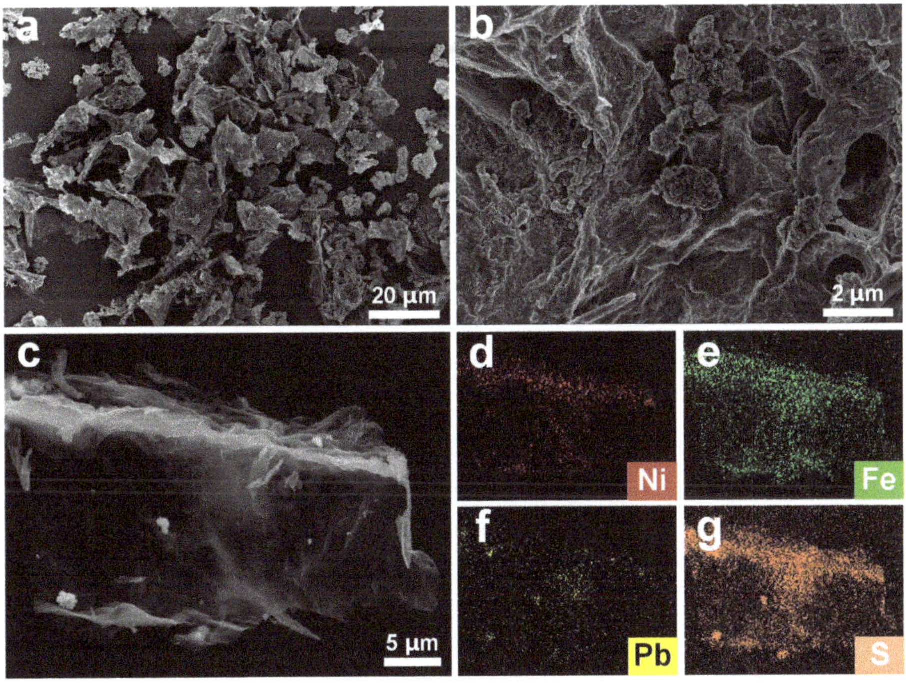

Figure 4. (**a**) SEM images of NiFe-PbO_2/S/rGO-45, (**b**) SEM images of NiFe-PbO_2/S/rGO-45, SEM mapping area of (**c**) NiFe-PbO_2/S/rGO-45, and the elemental distribution of (**d**) Ni, (**e**) Fe, (**f**) Pb, (**g**) S.

EDS mapping further indicates that Ni, Fe, Pb, and S are homogeneously distributed across the rGO surface (Figures 4c–g and S1). NiFe LDH and PbO_2 are loaded onto rGO as nanoparticles (NiFe-PbO_2/S/rGO-45), which significantly enhances the material's electrocatalytic performance. The nanoparticle diameter is approximately 29 nm (Figure S2), and their distribution on the rGO surface facilitates water molecule adsorption and accelerates electron transfer, further improving the activity.

XPS was used to further examine the impact of plasma treatment on the chemical states of NiFe-PbO_2/S/rGO-X. Figure S3 illustrates that the detection of N 1s and S 2p peaks in the XPS full spectrum of NiFe-PbO_2/S/rGO-45 signifies that high-energy plasma activates anionic precursors, producing active free radicals that enhance element doping. Plasma-induced groups specifically degrade thiourea, yielding the intermediate product S^{2-}. The sulfur species are then stimulated or ionized in the plasma, resulting in the integration of free sulfur atoms into the graphene structure. The C 1s fine spectra (Figure 5a) display 284.8 eV, 285.7 eV, and 288.9 eV, which are attributed to C-C, C-S, and C=O bonds, respectively. In comparison to the C 1s fine spectrum of GO of comparable purity documented in the literature, the strength of the C=O peak in NiFe-PbO_2/S/rGO-45 is markedly diminished, thereby validating the effective reduction of GO to rGO through the elimination of oxygen species [30,31]. The peak fitting of the Ni 2p fine spectrum (Figure 5b) identifies four unique chemical states of Ni. 856.7 eV and 874.5 eV correspond to the Ni $2p_{3/2}$ and Ni $2p_{1/2}$ orbitals linked to Ni^{3+} ions. Ni^{3+} ions can function as p-type dopants by introducing holes or as n-type dopants by introducing additional electrons. 861.7 eV and 880.4 eV are satellite peaks [32,33]. The Fe 2p spectrum (Figure 5c) exhibits 713.1 eV and 726.3 eV, which correspond to Fe^{3+} ions, signifying the oxidation of Fe^{2+} to Fe^{3+} during plasma treatment [34]. The high-resolution Pb 4f spectrum (Figure 5d) has prominent peaks at 137 eV and 141.8 eV, indicative of Pb^{4+}, and at 138.1 eV and 143 eV, representative of Pb^{2+}. This suggests that the Pb ions in NiFe-PbO_2/S/rGO-45 exist in a mixed valence state of Pb^{2+} and Pb^{4+}. The presence of Pb vacancies adjacent to Pb ions provides additional electrons to the O ions, which have detached from the Pb, thereby achieving charge balance and enhancing the catalytic properties [35].

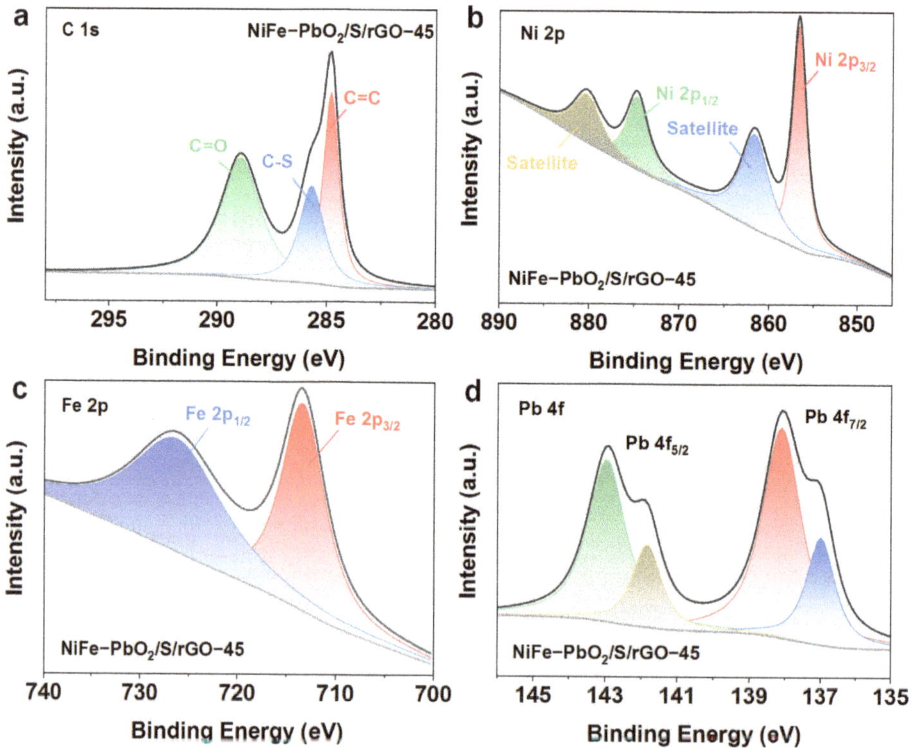

Figure 5. XPS spectrum of NiFe-PbO_2/S/rGO-45 (**a**) Ni 2p, (**b**) Fe 2p, (**c**) Pb 4f, (**d**) S 2p.

The OER activity of NiFe-PbO_2/S/rGO-X was assessed utilizing a three-electrode setup in a 1 mol·L^{-1} KOH, with commercial IrO_2 examined under identical circumstances for comparative analysis. As shown in Figure 6a, the LSV polarization curves reveal that NiFe-PbO_2/S/rGO-45 exhibits the lowest overpotential, with an η_{50} of 230 mV, significantly lower than the commercial catalyst IrO_2 (η_{50} for 50 mA·cm^{-2} overpotential), and superior to similar OER catalysts reported in recent literature (Figure 6c) [36–41]. However, as the plasma treatment time increased, the overpotential did not continue to decrease, likely due to the excessive formation of vacancy defects, which could lead to structural instability of the catalyst. In Figure 6b, the Tafel slope of NiFe-PbO_2/S/rGO-45 (44.26 mV·dec^{-1}) is the lowest, indicating fast reaction kinetics. This is further supported by the EIS results (Figure S4), which show that NiFe-PbO_2/S/rGO-45 has the lowest charge transfer impedance. The enhanced conductivity and electron transfer rates are attributed to the introduction of PbO_2 and the vacancy-defect graphene structure. These structural features contribute to improved reaction kinetics and optimized electron distribution, making the catalyst highly efficient for OER. The activation reaction surface area of NiFe-PbO_2/S/rGO-45, as shown in Figure S5, is the largest, indicating that the introduction of graphene and vacancy defects increases the catalyst's reactive surface area. This indicates that vacancy defects and the graphene matrix collectively significantly enhance the catalyst's performance.

Furthermore, following 50 h of continuous potential assessment, the current density of NiFe-PbO_2/S/rGO-45 exhibited minimal decline (Figure 6d), demonstrating its excellent stability. This suggests that the vacancy-defect structure not only improves the catalyst's reaction kinetics but also contributes to its long-term durability.

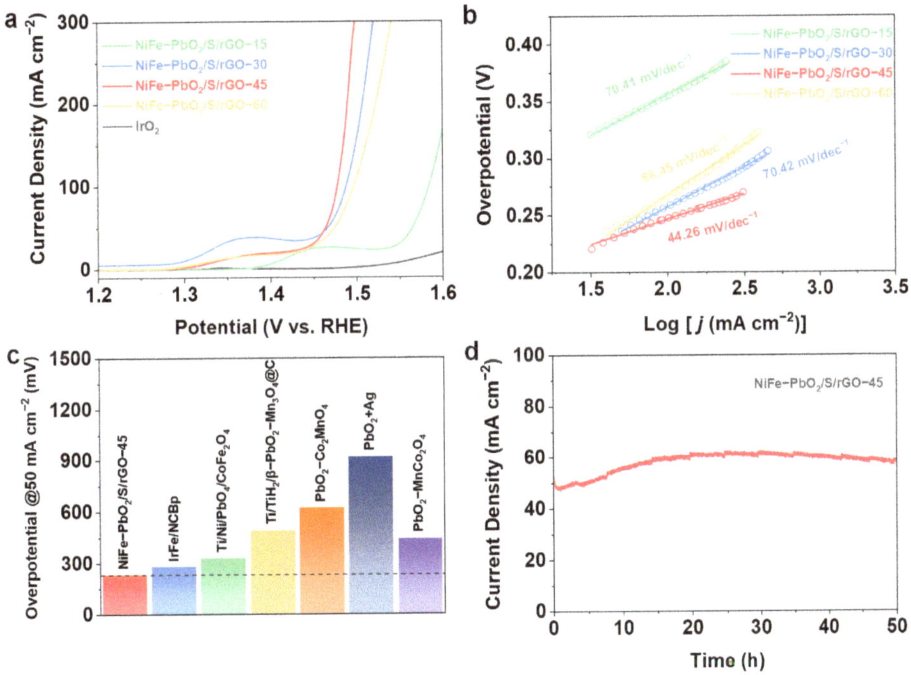

Figure 6. OER performances of NiFe-PbO_2/S/rGO-45. (**a**) LSV curve, (**b**) Tafel slope, (**c**) Comparison of η_{50}. (**d**) Amperometric i–t curve.

To further verify the exceptional activity of NiFe-PbO_2/S/rGO-45 and explore the changes during the OER process, XRD characterization was performed on the samples before and after 50 h of potentiostatic testing (Figure S6). The results show a small amount of Fe_3O_4 formation after the reaction, which likely exposes iron active sites and facilitates water molecule adsorption. SEM observations (Figure S7) reveal surface degradation and nanoparticle aggregation post-OER testing, indicating structural changes after prolonged testing. ICP-MS analysis (Figure S8) of the concentrations of Ni, Fe, Pb, and S during the CV activation process reveals that Ni, Fe, and Pb dissolve and redeposit in the electrolyte, while sulfur dissolves only slightly and undergoes a gradual dissolution process, which is beneficial for the sample's stability.

Building on these experimental findings, DFT calculations were performed to assess the impact of defect structures on catalytic performance. Based on the XRD analysis, the (003) crystal plane, most exposed to NiFe-LDH, was chosen for modeling. The NiFe-PbO_2/S/V_G and V_G heterojunction models, which include abundant defect structures (with vacancies, V, and graphene, G), were constructed for the typical OER reaction model [42]. The Gibbs free energy curve demonstrates that the introduction of vacancies in graphene reduces ΔG_{OH} and lowers overpotential. The presence of sulfur vacancies improves adsorption efficiency, while the addition of oxygen further stabilizes the structure, altering the rate-determining step of the OER from ΔG_{OH} in V_G to ΔG_{OOH} in NiFe-PbO_2/S/V_G. The overpotential of the rate-determining step for NiFe-PbO_2/S/V_G is 0.46 eV (Figure 7a).

Vacancies substantially influence the electrical structure and conductivity, optimizing the adsorption of OER intermediates, hence improving total activity. Vacancy structures create extra active sites, enhancing the adsorption and reactivity of intermediates, while alterations in electronic structure improve charge transfer kinetics, hence augmenting OER performance [43]. In addition, NiFe-PbO_2/S/V_G exhibits good thermal stability as demonstrated by small total energy and temperature fluctuations in AIMD simulations (Figures 7b and S9). Furthermore, the material maintains its structural integrity, with stable vacancy configurations, contributing to its overall robustness.

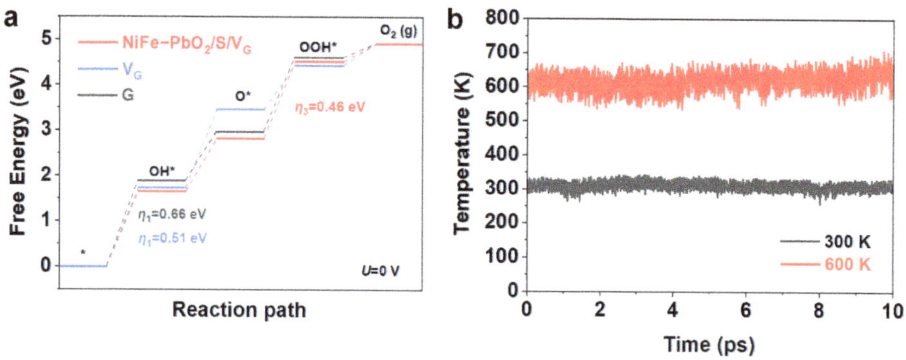

Figure 7. (**a**) OER free-energy diagram of NiFe-PbO_2/S/V_G and V_G, (**b**) Temperature fluctuation for NiFe-PbO_2/S/V_G.

4. Conclusions

In summary, NiFe-PbO_2/S/rGO-45, a newly conceived 2D material system, was manufactured via RF plasma, allowing for meticulous control over the material's characteristics. The integration of a graphene framework improved the catalyst's conductivity, while the plasma etching process reduced GO to rGO, facilitating the loading of NiFe LDH and PbO_2 as well as sulfur doping. This plasma-assisted reduction also generated carbon defects and sulfur vacancies, which promote charge transfer, and enhance the distribution of active sites, collectively accelerating the surface OER process. As a result, NiFe-PbO_2/S/rGO-45 exhibited superior OER performance compared to commercial IrO_2, with a η_{50} of 230 mV. The novelty of this material system lies in the combination of the graphene framework and the layered structures of NiFe-LDH and PbO_2, enhancing the catalyst's performance. In addition, the RF plasma-assisted synthesis method is scalable and adaptable to industrial-scale manufacturing processes. The use of abundant and inexpensive transition metals, combined with the graphene framework, makes the NiFe-PbO_2/S/rGO material system an economically viable alternative to precious metal-based catalysts. The superior OER performance demonstrated has the potential to greatly enhance the efficiency and cost-effectiveness of hydrogen production through water electrolysis, contributing to the global efforts in transitioning towards a sustainable and low-carbon energy future.

Supplementary Materials: The following supporting information can be downloaded at https://www.mdpi.com/article/10.3390/ma18010121/s1, Figure S1: EDS pattern of NiFe-PbO_2/S/rGO-45. Figure S2: Particle size of NiFe-PbO_2/S/rGO-45. Figure S3: XPS survey spectrum of NiFe-PbO_2/S/rGO-45. Figure S4: Nyquist plot of NiFe-PbO_2/S/rGO-X. Figure S5: Capacitive current density against scan rates of NiFe-PbO_2/S/rGO-X. Figure S6: XRD patterns of NiFe-PbO_2/S/rGO-45 before reaction, after OER processes in 1 M KOH for 50 h. Figure S7: SEM image of NiFe-PbO_2/S/rGO-45 before reaction, after OER for 50 h. Figure S8: Concentrations of Ni, Fe, Pb, and S in the electrolyte of NiFe-PbO_2/S/rGO-45 prior to the reaction and subsequent to the OER processes in 1 M KOH for 50 h, as determined by ICP-MS analysis. Figure S9: Energy fluctuation for NiFe-$PbO_2/S/V_G$ at (a) 600 K and (b) 300 K.

Author Contributions: Conceptualization, T.Y.; Funding acquisition, Z.Z.; Investigation, T.Y. and Z.Z.; Methodology, T.Y.; Project administration, H.L., X.L. and Q.Y.; Resources, T.Y. and Z.Z.; Supervision, H.L., X.L. and Q.Y.; Writing—original draft, F.T. and H.W.; Writing—review and editing, T.Y. and H.L. All authors have read and agreed to the published version of the manuscript.

Funding: This research was funded by the State Grid Chongqing Electric Power Company Technology Project (No. 522029240001).

Institutional Review Board Statement: Not applicable.

Informed Consent Statement: Not applicable.

Data Availability Statement: Data are contained within the article or Supplementary Material.

Conflicts of Interest: Authors Tingting Yang, Zheng Zhang, Fei Tan, Huayu Liu, and Xingyu Li were employed by the company State Grid Chongqing Electric Power Company Material Branch. Author Hongqi Wang was employed by the company Chongqing Jie Chuang Electric Power Technology Company Ltd. The remaining authors declare that the research was conducted in the absence of any commercial or financial relationships that could be construed as potential conflicts of interest.

References

1. Jawhari, A.H.; Hasan, N. Nanocomposite Electrocatalysts for Hydrogen Evolution Reactions (HERs) for Sustainable and Efficient Hydrogen Energy-Future Prospects. *Materials* **2023**, *16*, 3760. [CrossRef]
2. Popov, A.A.; Afonnikova, S.D.; Varygin, A.D.; Bauman, Y.I.; Trenikhin, M.V.; Plyusnin, P.E.; Shubin, Y.V.; Vedyagin, A.A.; Mishakov, I.V. Pt1-xNix Alloy Nanoparticles Embedded in Self-Grown Carbon Nanofibers: Synthesis, Properties and Catalytic Activity in HER. *Catalysts* **2023**, *13*, 599. [CrossRef]
3. Shamskhou, K.; Awada, H.; Yari, F.; Aljabour, A.; Schöfberger, W. A Molecular Binuclear Nickel (II) Schiff Base Complex for Efficient HER Electrocatalysis. *Catalysts* **2023**, *13*, 1348. [CrossRef]
4. Seenivasan, S.; Seo, J. Inverting destructive electrochemical reconstruction of niobium nitride catalyst to construct highly efficient HER/OER catalyst. *Chem. Eng. J.* **2023**, *454*, 140558. [CrossRef]
5. Xin, Y.M.; Hua, Q.Q.; Li, C.J.; Zhu, H.D.; Gao, L.G.; Ren, X.F.; Yang, P.X.; Liu, A.M. Enhancing electrochemical performance and corrosion resistance of nickel-based catalysts in seawater electrolysis: Focusing on OER and HER. *J. Mater. Chem. A* **2024**, *12*, 23147–23178. [CrossRef]
6. Gan, J.C.; Jiang, Z.F.; Fang, K.M.; Li, X.S.; Zhang, L.; Feng, J.J.; Wang, A.J. Low Rh doping accelerated HER/OER bifunctional catalytic activities of nanoflower-like Ni-Co sulfide for greatly boosting overall water splitting. *J. Colloid Interface Sci.* **2025**, *677*, 221–231. [CrossRef]
7. Khan, I.; Baig, N.; Bake, A.; Haroon, M.; Ashraf, M.; Al-Saadi, A.; Tahir, M.N.; Wooh, S. Robust electrocatalysts decorated three-dimensional laser-induced graphene for selective alkaline OER and HER. *Carbon* **2023**, *213*, 118292. [CrossRef]
8. Liao, Y.Y.; He, R.C.; Pan, W.H.; Li, Y.; Wang, Y.Y.; Li, J.; Li, Y.X. Lattice distortion induced Ce-doped NiFe-LDH for efficient oxygen evolution. *Chem. Eng. J.* **2023**, *464*, 142669. [CrossRef]
9. Shi, J.W.; He, H.W.; Guo, Y.H.; Ji, F.; Li, J.; Zhang, Y.; Deng, C.W.; Fan, L.Y.; Cai, W.W. Enabling high-efficiency ethanol oxidation on NiFe-LDH via deprotonation promotion and absorption inhibition. *J. Energy Chem.* **2023**, *85*, 76–82. [CrossRef]
10. Li, M.Y.; Yang, J.K.; Liang, S.; Hou, H.J.; Hu, J.P.; Liu, B.C.; Kumar, R.V. Review on clean recovery of discarded/spent lead-acid battery and trends of recycled products. *J. Power Sources* **2019**, *436*, 226853. [CrossRef]
11. Hakimi, F.; Ghalkhani, M.; Rashchi, F.; Dolati, A. Pulse electrodeposition synthesis of Ti/PbO_2-IrO_2 nano-composite electrode to restrict the OER in the zinc electrowinning. *J. Environ. Chem. Eng.* **2024**, *12*, 111985. [CrossRef]
12. Lin, L.; Wang, Y.F.; Ye, Q.; Zhao, Y.X.; Cheng, Y.L. Rapid fabrication of $Fe_xNi_{2-x}P_4O_{12}$ and graphene hybrids as electrocatalyst for highly efficient oxygen evolution reaction. *Appl. Catal. B-Environ. Energy* **2023**, *334*, 122834. [CrossRef]
13. González-Ingelmo, M.; García, M.L.; Oropeza, F.E.; Alvarez, P.; Blanco, C.; Santamaría, R.; Rocha, V.G. Ultra-high dispersion of Ni-based OER catalysts on graphene 3D networks enhances the in situ Fe^{3+} catalytic activation. *J. Mater. Chem. A* **2023**, *11*, 24248–24260. [CrossRef]
14. Gong, Z.C.; Liu, J.J.; Yan, M.M.; Gong, H.S.; Ye, G.L.; Fei, H.L. Highly Durable and Efficient Seawater Electrolysis Enabled by Defective Graphene-Confined Nanoreactor. *ACS Nano* **2023**, *17*, 18372–18381. [CrossRef] [PubMed]

15. Goyenola, C.; Stafström, S.; Hultman, L.; Gueorguiev, G.K. Structural Patterns Arising during Synthetic Growth of Fullerene-Like Sulfocarbide. *J. Phys. Chem. C* **2012**, *116*, 21124–21131. [CrossRef]
16. Sfuncia, G.; Nicotra, G.; Giannazzo, F.; Pécz, B.; Gueorguiev, G.K.; Kakanakova-Georgieva, A. 2D graphitic-like gallium nitride and other structural selectivity in confinement at the graphene/SiC interface. *Crystengcomm* **2023**, *25*, 5810–5817. [CrossRef]
17. Yang, H.C.; Wang, C.H.; Zhang, Y.J.; Wang, Q.B. Green synthesis of NiFe LDH/Ni foam at room temperature for highly efficient electrocatalytic oxygen evolution reaction. *Sci. China-Mater.* **2019**, *62*, 681–689. [CrossRef]
18. Hong, X.D.; Li, S.L.; Tang, X.N.; Sun, Z.H.; Li, F. Self-supporting porous CoS_2/rGO sulfur host prepared by bottom-up assembly for lithium-sulfur batteries. *J. Alloy. Compd.* **2018**, *749*, 586–593. [CrossRef]
19. Pimenta, M.A.; Dresselhaus, G.; Dresselhaus, M.S.; Cançado, L.G.; Jorio, A.; Saito, R. Studying disorder in graphite-based systems by Raman spectroscopy. *Phys. Chem. Chem. Phys.* **2007**, *9*, 1276–1291. [CrossRef]
20. Zhang, Y.Q.; Liang, Y.M.; Zhou, J.X. Recent Progress of Graphene Doping. *Acta Chim. Sin.* **2014**, *72*, 367–377. [CrossRef]
21. Zhu, J.W.; Mu, S.C. Defect Engineering in the Carbon-Based Electrocatalysts: Insight into the Intrinsic Carbon Defects. *Adv. Funct. Mater.* **2020**, *30*, 2001097. [CrossRef]
22. Huang, X.X.; Shen, T.; Zhang, T.; Qiu, H.L.; Gu, X.X.; Ali, Z.; Hou, Y.L. Efficient Oxygen Reduction Catalysts of Porous Carbon Nanostructures Decorated with Transition Metal Species. *Adv. Energy Mater.* **2020**, *10*, 1900375. [CrossRef]
23. Kim, J.H.; Choi, K.J.; Yoon, S.G. Electrical and reliability characteristics of HfO_2 gate dielectric treated in N_2 and NH_3 plasma atmosphere. *Appl. Surf. Sci.* **2005**, *242*, 313–317. [CrossRef]
24. Petrovic, M.; Jovanovic, T.; Rancev, S.; Kovac, J.; Velinov, N.; Najdanovic, S.; Kostic, M.; Bojic, A. Plasma modified electrosynthesized cerium oxide catalyst for plasma and photocatalytic degradation of RB 19 dye. *J. Environ. Chem. Eng.* **2022**, *10*, 107931. [CrossRef]
25. Rousseau, A.; Guaitella, O.; Gatilova, L.; Thevenet, F.; Guillard, C.; Röpcke, J.; Stancu, G.D. Photocatalyst activation in a pulsed low pressure discharge. *Appl. Phys. Lett.* **2005**, *87*, 221501. [CrossRef]
26. Czech, T.; Sobczyk, A.T.; Jaworek, A. Optical emission spectroscopy of point-plane corona and back-corona discharges in air. *Eur. Phys. J. D* **2011**, *65*, 459–474. [CrossRef]
27. Lu, L.L.; Xu, X.X.; An, K.L.; Wang, Y.; Shi, F.N. Coordination Polymer Derived NiS@g-C_3N_4 Composite Photocatalyst for Sulfur Vacancy and Photothermal Effect Synergistic Enhanced H2 Production. *ACS Sustain. Chem. Eng.* **2018**, *6*, 11869–11876. [CrossRef]
28. Xie, J.F.; Zhang, H.; Li, S.; Wang, R.X.; Sun, X.; Zhou, M.; Zhou, J.F.; Lou, X.W.; Xie, Y. Defect-Rich MoS_2 Ultrathin Nanosheets with Additional Active Edge Sites for Enhanced Electrocatalytic Hydrogen Evolution. *Adv. Mater.* **2013**, *25*, 5807–5813. [CrossRef] [PubMed]
29. Li, H.; Tsai, C.; Koh, A.L.; Cai, L.L.; Contryman, A.W.; Fragapane, A.H.; Zhao, J.H.; Han, H.S.; Manoharan, H.C.; Abild-Pedersen, F.; et al. Activating and optimizing MoS_2 basal planes for hydrogen evolution through the formation of strained sulphur vacancies. *Nat. Mater.* **2016**, *15*, 48–53. [CrossRef]
30. Bo, L.L.; Shi, W.P.; Nian, F.; Hu, Y.S.; Pu, L.M.; Li, P.; Zhang, Z.X.; Tong, J.H. Interface engineering of Co_9S_8@Co_3O_4/N, S-doped carbon core@shell nanostructures serve as an excellent bifunctional ORR/OER electrocatalyst for rechargeable Zn-air battery. *Sep. Purif. Technol.* **2023**, *307*, 122536. [CrossRef]
31. Chen, D.L.; Gan, C.L.; Fan, X.Q.; Zhang, L.; Li, W.; Zhu, M.H.; Quan, X. Improving the Dynamic Mechanical Properties of XNBR Using ILs/KH550-Functionalized Multilayer Graphene. *Materials* **2019**, *12*, 2800. [CrossRef] [PubMed]
32. Wen, P.; Lei, R.B.; Cao, X.; Ma, Q.; Zhang, G.W.; Guo, C.X.; Wang, X.W.; Qiu, Y.J. Anchored Ni nanocrystals boosting $BiVO_4$ photoanode for highly efficient water oxidation via in-situ generation of Ni@NiOOH co-catalyst. *Chem. Eng. J.* **2023**, *454*, 139983. [CrossRef]
33. Chen, Q.; Huang, J.F.; Xiao, T.; Cao, L.Y.; Liu, D.H.; Li, X.Y.; Niu, M.F.; Xu, G.T.; Kajiyoshi, K.; Feng, L.L. V-doped Ni_2P nanoparticle grafted g-C_3N_4 nanosheets for enhanced photocatalytic hydrogen evolution performance under visible light. *Dalton Trans.* **2023**, *52*, 7447–7456. [PubMed]
34. Ye, X.M.; Meng, X.N.; Han, Z.Q.; Qi, Y.G.; Li, Z.M.; Tian, P.P.; Wang, W.S.; Li, J.; Li, Y.C.; Zhang, W.C.; et al. Designing Fe-containing polyhedral oligomeric silsesquioxane to endow superior mechanical and flame-retardant performances of polyamide 1010. *Compos. Sci. Technol.* **2023**, *233*, 109894. [CrossRef]
35. Jiang, Z.X.; Yu, T.Q.; Chen, J.L.; Tan, K.X.; Deng, R.; Zhou, A.C.; Yin, S.B. Regulating Competitive Adsorption on Pt Nanoparticles by Introducing Pb to Expedite Hydrogen Production via Ammonia Oxidation. *ACS Appl. Nano Mater.* **2023**, *6*, 1889–1897. [CrossRef]
36. Hamze, M.; Rezaei, M.; Tabaian, S.H. Cobalt ferrite coated on Ti/Ni/PbO_x with enhanced electrocatalytic stability for chloride ions-contained water splitting. *J. Electroanal. Chem.* **2023**, *948*, 117823. [CrossRef]
37. Jiang, G.S.; Chen, M.L.; Sun, Y.Z.; Wu, Y.F.; Pan, J.Q. Highly dispersed Ir/Fe nanoparticles anchored at nitrogen-doped activated pyrolytic carbon black as a high-performance OER catalyst for lead recovery. *Dalton Trans.* **2024**, *53*, 1121–1131. [CrossRef] [PubMed]

38. Li, S.M.; Shi, M.; Wu, C.X.; Nie, K.Q.; Wei, Z.; Jiang, X.P.; Liu, X.B.; Chen, H.L.; Tian, X.L.; Wu, D.X.; et al. Surface addition of Ag on PbO_2 to enable efficient oxygen evolution reaction in pH-neutral media. *Chem. Eng. J.* **2024**, *485*, 150043. [CrossRef]
39. Wang, X.B.; Wang, J.L.; Jiang, W.H.; Chen, C.; Wei, J.L.; Yu, B.H.; Chen, B.M.; Xu, R.D.; Yang, L.J. $MnCo_2O_4$ decorating porous PbO2 composite with enhanced activity and durability for acidic water oxidation. *Fuel* **2023**, *338*, 127344. [CrossRef]
40. Wang, X.B.; Wang, J.L.; Tong, X.N.; Wu, S.; Wei, J.L.; Chen, B.M.; Xu, R.D.; Yang, L.J. Constructing of Pb-Sn/a-PbO_2/β-PbO_2-Co_2MnO_4 composite electrode for enhanced oxygen evolution and zinc electrowinning. *Mater. Today Phys.* **2023**, *35*, 101068.
41. Wu, S.; Wang, J.L.; Wang, X.B.; Jiang, D.; Wei, J.L.; Tong, X.N.; Liu, Z.W.; Kong, Q.X.; Zong, N.X.; Xu, R.D.; et al. Mn_3O_4@C micro-flakes modified Ti/TiH_2/β-PbO_2 anode for accelerating oxygen evolution reaction in zinc electrowinning. *Mater. Res. Bull.* **2024**, *171*, 112605. [CrossRef]
42. Bajdich, M.; García-Mota, M.; Vojvodic, A.; Norskov, J.K.; Bell, A.T. Theoretical Investigation of the Activity of Cobalt Oxides for the Electrochemical Oxidation of Water. *J. Am. Chem. Soc.* **2013**, *135*, 13521–13530. [CrossRef] [PubMed]
43. Li, X.M.; Zheng, K.T.; Zhang, J.J.; Li, G.N.; Xu, C.J. Engineering Sulfur Vacancies in Spinel-Phase Co3S4 for Effective Electrocatalysis of the Oxygen Evolution Reaction. *ACS Omega* **2022**, *7*, 12430–12441. [CrossRef] [PubMed]

materials

Article

XPS Depth Profiling of Surface Restructuring Responsible for Hydrogen Evolution Reaction Activity of Nickel Sulfides in Alkaline Electrolyte

Jiangtian Li [1,*], Deryn Chu [1,*], Connor Poland [2], Cooper Smith [2], Enoch A. Nagelli [2] and Victor Jaffett [2]

1 U.S. Army DEVCOM Army Research Laboratory, 2800 Powder Mill Rd, Adelphi, MD 20783, USA
2 Department of Chemistry & Life Science, United States Military Academy, West Point, NY 10996, USA; connor.poland@westpoint.edu (C.P.); cooper.smith@westpoint.edu (C.S.); enoch.nagelli@westpoint.edu (E.A.N.); victor.jaffett@westpoint.edu (V.J.)
* Correspondence: jiangtian.li.ctr@army.mil (J.L.); deryn.d.chu.civ@army.mil (D.C.)

Abstract: Electrochemical water splitting provides a sustainable method for hydrogen production. However, the primary challenge for electrochemical hydrogen generation is the high cost and limited availability of platinum-based noble-metal catalysts. Transition-metal chalcogenides have been identified as low-cost and efficient electrocatalysts to promote the hydrogen evolution reaction (HER) in alkaline electrolytes. Nonetheless, the identification of active sites and the underlying catalytic mechanism remain elusive. In this study, phosphorus-doped nickel sulfide has been successfully synthesized, demonstrating enhanced activity for alkaline HER. Investigating surface chemistry through X-ray photoelectron spectroscopy (XPS), depth profiling revealed that surface restructuring occurs during the HER process. The presence of phosphorus significantly influences this transformation, promoting the formation of a novel active Ni-O layer. This Ni-O layer is responsible for enhanced catalytic activity by upshifting the *d*-band center and increasing the density of states near the Fermi level, along with expanding the electrochemical surface area. This study reveals that the surface restructuring of transition-metal sulfides is highly tied to the electronic structure of the parent catalysts. Gaining a comprehensive understanding of this surface restructuring is essential for predicting and exploring more efficient non-precious transition-metal sulfide electrocatalysts.

Keywords: water splitting; hydrogen evolution; sulfides; surface restructuring; XPS depth profile

1. Introduction

Hydrogen has emerged as a green energy carrier due to its high energy density and environmentally friendly properties. However, high-purity hydrogen generation, storage, and transportation have long been significant bottlenecks to the development and practical implementation of hydrogen fuel cell technologies [1–10]. Renewable electricity-powered water electrolysis to produce hydrogen is broadly recognized as a viable technology for a clean and sustainable global energy future [4–10], especially when considered to be integrated with renewable energy capture and onsite fuel cell systems. The key to enabling this technology lies in the development of low-cost and highly efficient electrocatalysts to accelerate water-splitting reactions. In particular, designing high-performance electrocatalysts for HER under alkaline conditions is more practical, as alkaline water electrolysis is the most commonly used approach in industry [10]. Nonetheless, alkaline HER suffers

https://doi.org/10.3390/ma18030549

from more sluggish reaction kinetics compared to acidic electrolysis due to requiring an initial step to dissociate water molecules and thereafter create protons (H^+) for subsequent H_2 generation ($H_2O \rightarrow H^+ + OH^-$) [9–17]. To date, platinum-based precious materials are acknowledged as the benchmark electrocatalysts for HER thanks to their optimal adsorption/desorption energy with protons. Nonetheless, the widespread application of noble-metal-based electrocatalysts is dramatically hindered by their scarcity, leading to high costs. As a result, considerable efforts have been made to identify high-performance, earth-abundant, and cost-effective non-precious HER electrocatalysts [3,10,11,17].

In the last decade, transition-metal sulfides have been regarded as promising alternatives to platinum-group catalysts. Notably, nickel sulfides stand out for their excellent electrical conductivity, high redox current responses, and cost-effectiveness [18–25]. Ni_3S_2, a metal chalcogenide naturally occurring as the mineral heazlewoodite, exhibits intrinsic metallic behavior due to a continuous network of short Ni-Ni bonds throughout its structure [18]. In addition, Ni_3S_2 is the most nickel-rich phase in the stoichiometric formula, offering abundant Ni active sites [19,24]. Its high conductivity, coupled with its low cost, makes Ni_3S_2 well-suited for various electrochemical applications. While recently Ni_3S_2 based materials have been shown to be electrocatalytic active for either HER or OER, their catalytic activity and stability are still less competitive compared to those of the noble-metal catalysts.

Fundamentally, fostering catalytic activity in water splitting relies on optimizing the chemisorption of reactants and intermediates at the active sites. A variety of strategies have been employed to manipulate the electronic structures of nickel sulfides, aiming to regulate the adsorption behavior of intermediates on active sites and enhance HER performance. These strategies include the incorporation of foreign atoms, morphological engineering, and the construction of heterojunctions [18–33]. The formation of heterojunctions between Ni_3S_2 and either Ni or MoO_2, for example, facilitated the adsorption and dissociation of water molecules into hydroxide ions (OH^-) and protons (H^+), thereby accelerating the hydrogen production process [20]. Additionally, the incorporation of the second cation (Fe, W, Sn, Mo, etc.) or ion (phosphorus, fluorine, and nitrogen) into sulfides has been revealed to enhance charge-transfer rates and increase the electrochemical surface area, resulting in improved HER activity [22,24–31]. Despite these significant efforts, the water-splitting performance of nickel sulfides remains inferior to that of noble-metal-based catalysts, and the identification of active sites and the underlying catalytic mechanisms remains elusive. These conclusions are traditionally drawn under the assumption that no surface changes occur during the electrochemical reactions. However, surface restructuring has been observed in several electrochemical reactions, such as the oxygen evolution reaction involving metal sulfides, phosphides, and nitrides, where the top surface is completely transformed into oxy-hydroxyl oxides as the active species, along with ion leaching [3,6,8–13]. Under oxygen reduction reaction (ORR) conditions, Ni_3S_2 has been found to undergo surface reconstruction, forming a highly active, amorphous NiS phase approximately 2 nm in thickness [8]. In another study focused on alkaline HER, NiS was shown to undergo an in situ phase transition, resulting in an intimately mixed phase of Ni_3S_2 and NiO. This phase transformation generates highly active synergistic dual sites at the Ni_3S_2/NiO interface [10]. Our recent work also observed surface reconstruction in 3D NiOS nanostructures for alkaline HER, leading to the formation of a Ni-O catalyst layer functionalized with low-coordinated oxygen atoms and abundant vacancies. These features enable active sites for H_{ad} adsorption/desorption, while nearby $Ni^{\delta+}$ vacancies facilitate optimal interactions with OH_{ad} [9]. These studies highlight the dynamic nature of transition-metal chalcogenides in electrocatalysis [8–10]. Despite the observed surface restructuring, several important questions remain unanswered, such as: What parameters govern the reconstruction? How can surface restructuring be controlled or predicted? And how can restructuring be directly linked to improved catalytic performance? Surface restructuring remains challenging to fully characterize and control, and a more comprehensive understanding is required. This is highly dependent on the clear observation and analysis of the chemical compositions and states at the surface. Among a variety of techniques employed for analysis, XPS is a highly sensitive surface analysis method. When combined with sputtering to gradually remove material between analysis cycles without damaging the underlying material, depth profiling allows for high-resolution analysis of the atomic composition and chemical state of the electrode with depth [14–16]. The information gained from this technique could significantly enhance our understanding of the compositional and structural changes that occur during surface restructuring, as well as provide insights into how to control this process.

Traditionally, catalyst electrodes are prepared by depositing catalyst ink, mixed catalyst powders with binders (commonly the Nafion), on glassy carbon. The co-existence of organic binders will interfere with the in/ex-situ characterizations and thereby obscure our understanding of the intrinsic activity and the potential surface reconstruction of the catalysts. In this study, instead, pure nickel sulfides were successfully fabricated by direct sulfurization of Ni foam under hydrothermal conditions and subsequently doped with phosphorus, which is able to create a clean surface and allows us to examine surface changes after HER without interference from conventional binders. We evaluated the electrocatalytic HER performance of these catalysts in an alkaline electrolyte, finding significant enhancements due to phosphorus doping. XPS depth profiling of the post-HER electrodes revealed that surface restructuring occurred during the alkaline HER process for both sulfide electrodes. Additionally, the presence of phosphorus significantly impacts this restructuring process on the top layer by introducing crystal defects through the doping effect while protecting the further proceeding to the deeper layer. This behavior contributes to the robust stability of sulfide catalysts.

2. Materials and Methods

2.1. Fabrication of Ni_3S_2 Framework

All chemicals in this study were purchased from Sigma-Aldrich (MilliporeSigma, St. Louis, MO, USA) and used as received. The fabrication of Ni_3S_2 is technically based on the sulfurization of Ni foam with a sulfur source. Typically, 3.41 g of thiourea was dissolved in 90 mL of DI water in a Teflon liner, and then a piece of pre-cleaned Ni foam with a geometric size of 6 × 3.5 cm^2 was immersed in it. The Teflon liner was sealed in Autoclave and kept at 180 °C overnight. After it cooled down to room temperature, the reacted Ni foam was rinsed with DI water and ethanol and then dried under vacuum at 60 °C overnight.

2.2. Phosphorization of Ni_3S_2 (P-Ni_3S_2)

The phosphorization of Ni_3S_2 was carried out with NaH_2PO_4 as the phosphorous source. Typically, 1 g of NaH_2PO_4 in a ceramic boat was placed in a quartz tube, and one piece of Ni_3S_2 framework with a size of (3 × 3.5 cm^2) was placed on the downstream side. The tube furnace was flushed with N_2 for 30 min, then heated up to 300 °C with a ramping rate of 1 °C/min, and then kept for 2 h under N_2 flow with a flow rate of 100 SCCM (Standard Cubic Centimeters per Minute). The sample was denoted as P-Ni_3S_2.

2.3. Characterizations

The crystal structure was acquired on a Rigaku X-ray diffractometer (Rigaku Flex 600, Rigaku, Tokyo, Japan) with Cu Kα radiation (λ = 1.54 Å) within the diffraction angle range of 20 to 80°. Morphology of samples was observed using a Zeiss Auriga 60 Field

Emission Scanning Microscope (ZEISS, St. Louis, MO, USA), operating at an accelerating voltage of 10 keV. XPS measurements were conducted on a Physical Electronics VersaProbe III employing a monochromatic Al Kα source with an energy of 1486.6 eV. The analyzer pass energy was set to 224 eV for survey scans and 55 eV for high-resolution scans. All spectra were calibrated to the adventitious C 1s peak at 284.8 eV. For the XPS depth profile, all specimens were ion sputtered using an argon gun operating at 2 KeV over a 3 mm × 3 mm area for 5 cycles, with each cycle consisting of 4 min of sputtering and followed by data collection.

2.4. Electrochemical Performance

All electrochemical performance evaluations were carried out on a Gamry Reference 3000 (Gamry Instruments, Warminster, PA, USA) potentiostat. A three-electrode electrochemical cell was employed with the developed catalyst as the working electrode, the Ag/AgCl as the reference electrode, and the Pt wire as the counter electrode. 1 M KOH aqueous solution was used as the electrolyte during all electrochemical testing. Before conducting linear sweep voltammetry (LSV) with a scanning rate of 10 mV/s, chronopotentiometry was performed at −1.8 V vs. Ag/AgCl (−0.8 V vs. RHE) for 30 min to stabilize the current density. The potential vs. the Ag/AgCl electrode was calibrated to a reversible hydrogen electrode (RHE) with the following equation: E (RHE) = E (Ag/AgCl) + 0.059 pH + 0.197.

The electrochemical impedance spectroscopy (EIS) was conducted in a frequency range from 0.1 Hz to 100 kHz with an AC amplitude of 10 mV, and the electrocatalytic kinetics was studied by analyzing the semicircle of the Nyquist plot. The electrochemical double-layer capacitance (C_{dl}) was determined by running cyclic voltammetry (CV) in the non-Faradaic region with different scan rates. The electrochemically active surface area (ECSA) was then calculated using the following equation: ECSA = C_{dl}/C_s, where the C_s is the constant specific capacitance of the smoothy electrode (60 μF commonly used).

2.5. Preparation of Catalyst Ink and Pt Electrode

As a control sample for comparison, a Pt electrode was also prepared. A total of 5 mg of commercial Pt (20 wt.%)/C catalyst was dispersed in a mixture of 500 μL ethanol and 500 μL DI water with 10 μL of 5 wt.% 117 Nafion solution. The mixture was sonicated for 30 min to yield a homogeneous catalyst ink. Thereafter, 10 μL of ink was dropped on glassy carbon with a diameter of 3 mm. That will act as the Pt working electrode.

3. Results and Discussion

3.1. Catalyst Structure

To investigate the evolution of phase structure, the obtained samples were first characterized by X-ray diffraction (XRD), as shown in Figure 1. The Ni foam displays three distinct diffraction peaks (Figure 1a) corresponding to the (111), (200), and (220) planes of metal nickel (JCPDS No. 04-0850) [20]. In the sulfide sample, all diffraction peaks are indexed to the crystalline phase of heazlewoodite Ni_3S_2 (JCPDS No. 44-1418) [19,20], with no detectable peaks from impurities (Figure 1b), indicating a complete sulfurization of the Ni foam to Ni_3S_2 under this reaction condition. As shown in Figure 1c, the doping of phosphorus into the Ni_3S_2 lattice does not significantly alter the crystal structure of the parent material. However, due to the larger atomic radius of phosphorus compared to sulfur, the incorporation of phosphorus expands the crystal structure, leading to increased *d*-spacing and a redshift of the diffraction peaks towards smaller angles, as highlighted in the inset of Figure 1. Notably, no nickel phosphides were formed during the phosphorization process, confirming successful phosphorus doping in Ni_3S_2. The evolution of morphology was examined using scanning electron microscopy (SEM). Figure 2a,b present

SEM images at varying magnifications, illustrating that the Ni_3S_2 nanostructures effectively replicate the Ni foam skeleton without compromising its integrity during the sulfurization process. Notably, nanowires have grown on the surface, demonstrating excellent adhesion to the skeletal framework. This nanostructure offers distinct advantages over bulk materials, including a larger surface area and one-dimensional interwire channels, which facilitate electrolyte diffusion, enhance electron transport, and ensure ample contact between the active materials and the electrolyte [32]. The nanostructures were preserved in the phosphorus-doped catalyst P-Ni_3S_2, as shown in Figure 2c,d.

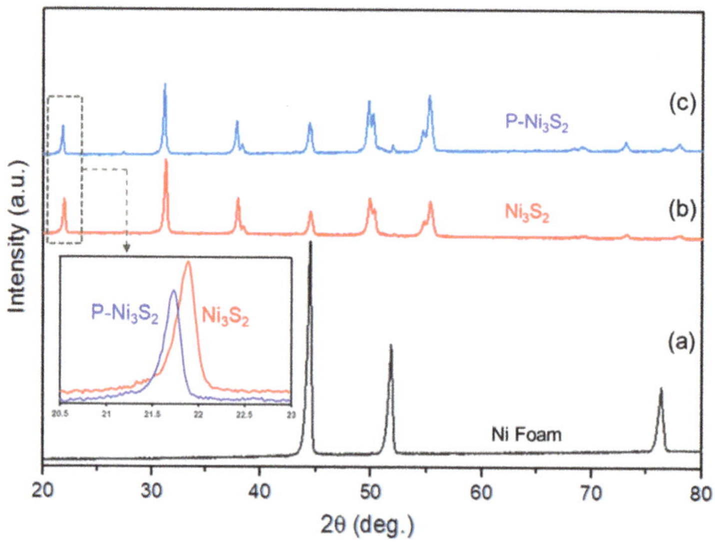

Figure 1. XRD patterns for Ni foam (**a**), Ni_3S_2 (**b**), and P-Ni_3S_2 (**c**). The inset compares the peak shift after P doping in Ni_3S_2.

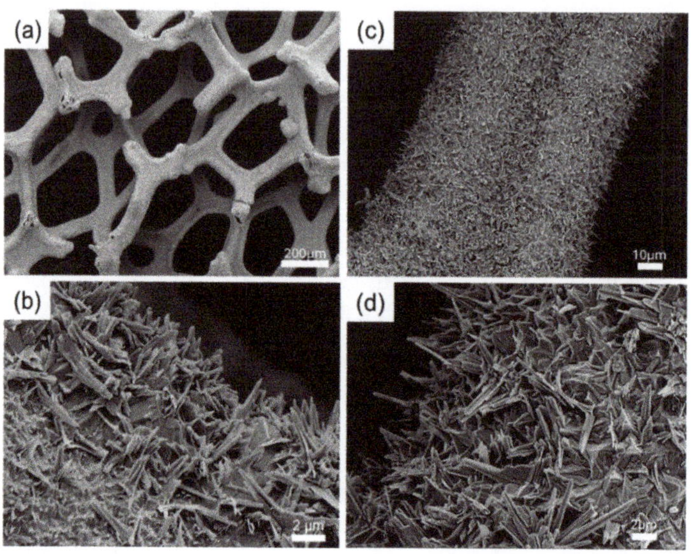

Figure 2. Surface morphology of developed catalysts. SEM images for freshly prepared Ni_3S_2 (**a**,**b**), and P-Ni_3S_2 (**c**,**d**), respectively, with different magnifications.

3.2. Surface Chemistry

An X-ray photoelectron spectroscopy (XPS) survey was conducted to verify the chemical composition of all catalysts, as shown in Figure 3. The predominant contributions were observed from C 1s, O 1s, Ni 2p, and Ni LMM. In the Ni_3S_2 sample, distinct sulfur features, S 2s and S 2p are clearly visible, while the phosphorus features, P 2s and P 2p, appear after phosphorization within the binding energy range of 100 to 250 eV. This indicates the successful incorporation of phosphorus.

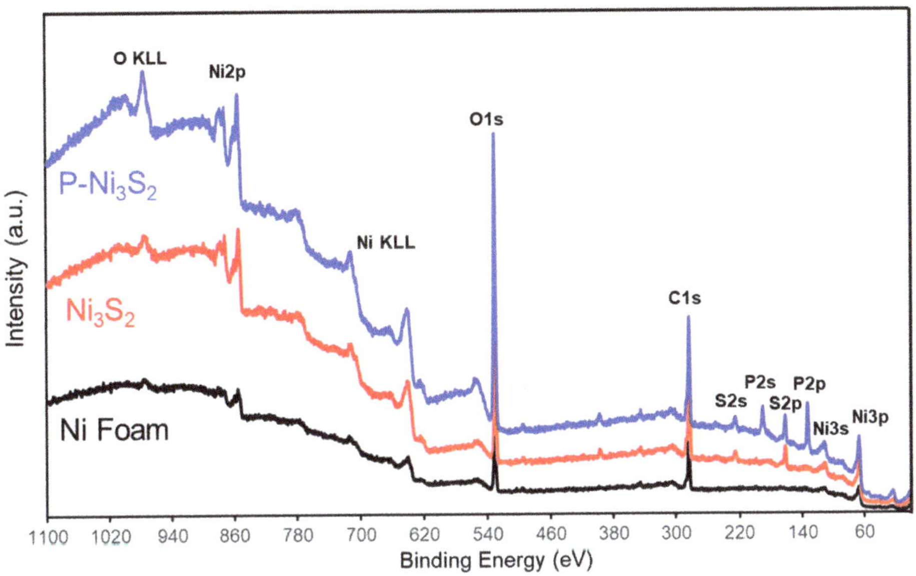

Figure 3. XPS surface survey of Ni foam, freshly prepared Ni_3S_2, and P-Ni_3S_2, respectively.

To closely examine variations in chemical compositions, high-resolution XPS spectra were obtained and are shown in Figure 4 for Ni 2p (a), S 2p (b), and P 2p (c), respectively. The successful doping of phosphorus is confirmed by the emergence of a P 2p peak at 135 eV in the P-Ni_3S_2 sample, in contrast to the flat signal observed in the Ni_3S_2 sample (Figure 4c). As shown in Figure 4a, both Ni 2p XPS spectra exhibit two main bands in the ranges of 850–865 eV and 870–885 eV, corresponding to the Ni $2p_{3/2}$ and Ni $2p_{1/2}$ spin-orbit doublets, respectively, with a splitting of approximately 17 eV and a Ni $2p_{3/2}$ to Ni $2p_{1/2}$ ratio of 2:1 [23,26,33–35]. Focusing on the $2p_{3/2}$ region in the pristine Ni_3S_2 sample, peaks at 855.3 eV and 852.7 eV are characteristic of Ni_3S_2 and are assigned to Ni-S and Ni-Ni bonds, respectively [23,35,36]. The bump centered around 861 eV arises from the satellite peak of Ni $2p_{3/2}$. In the P-Ni_3S_2 sample, two major changes in the Ni 2p spectra are observed following phosphorus doping: (i) the Ni-S peak shifts to a higher binding energy by 1.6 eV, reaching 856.9 eV, and the Ni-Ni peak shifts by 1.1 eV to 853.8 eV; (ii) the Ni-Ni ratio significantly decreases. In the S XPS spectra (Figure 4b), the major band centered around 163 eV is attributed to the S 2p signal, which can be deconvoluted into two peaks: S $2p_{3/2}$ and S $2p_{1/2}$. In the P-Ni_3S_2 sample, the S $2p_{3/2}$ peak shows a positive shift to 162.8 eV, an increase of 0.7 eV compared to Ni_3S_2 due to the presence of phosphorus. Overall, the incorporation of additional phosphorus ions into the structure extracts more electrons from the Ni atoms. As a result, some Ni-Ni bonds are partially broken and reformed with phosphorus, leading to the delocalization of electrons among the Ni cations and sulfur ions [20,37].

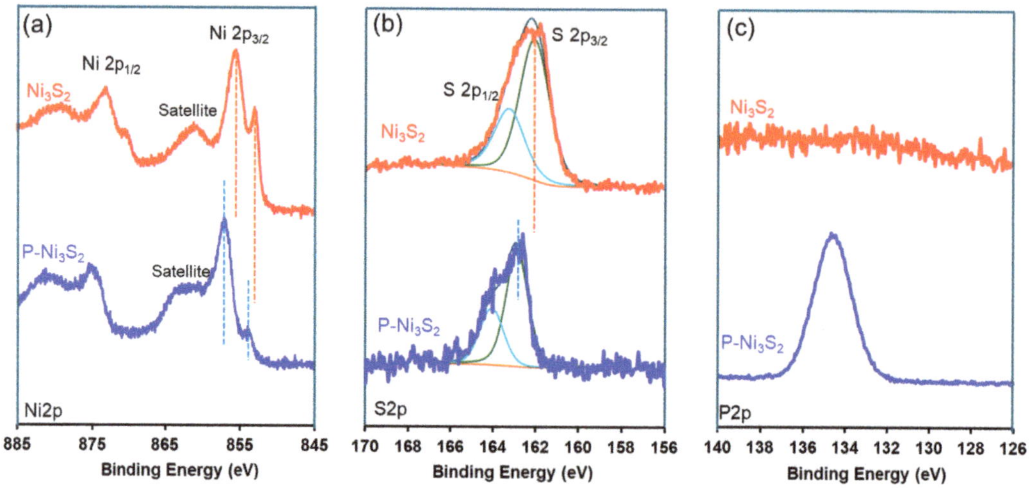

Figure 4. Surface chemistry of freshly prepared catalysts. High-resolution XPS spectra for Ni 2p (**a**), S 2p (**b**), and P 2p (**c**), respectively, in Ni_3S_2 (top) and P-Ni_3S_2 (bottom) prior to electrochemical testing.

3.3. Electrochemical Performances

The electrocatalytic activity of the developed catalysts toward HER under 1 M KOH alkaline electrolyte was evaluated using a typical three-electrode system, in which the catalyst was directly used as the working electrode. Figure 5a shows the polarization curves (linear sweep voltammetry, LSV) of NF, Ni_3S_2, P-Ni_3S_2, and Pt/C (20 wt.%), respectively. To achieve a current density of 10 mA/cm^2 for HER, NF requires an overpotential of 306 mV. In contrast, the corresponding HER overpotential for Ni_3S_2 is 181 mV, a substantial decrease of 125 mV compared to NF, indicating the significant activity of sulfides toward alkaline HER activity. A further remarkable improvement happens to the P-Ni_3S_2 catalyst with an overpotential of only 131 mV, a decrease of 50 mV owing to the phosphorus doping. For comparison, the Pt/C (20 wt.%) was also tested for alkaline HER under identical operation conditions. That demonstrates to be the most efficient catalyst with an overpotential of 48 mV at 10 mA/cm^2. To gain insight into the high HER activity for the P-Ni_3S_2 catalyst, the Tafel slopes were calculated, as shown in Figure 5b. The P-Ni_3S_2 catalyst reveals a Tafel slope of 84.6 mV/dec, much smaller than those of NF (120.7 mV/dec) and Ni_3S_2 catalyst (103.7 mV/dec), respectively. Fundamentally, hydrogen evolution in alkaline media occurs in two steps. The first step is the Volmer step, where the catalyst accepts one electron and cleaves the adsorbed H_2O molecule into a hydroxyl ion (OH^-) and an adsorbed hydrogen atom (H*), i.e., $H_2O + e^- + * \rightarrow H^* + OH^-$ (* denotes active site on the electrode surface). In the second step, hydrogen intermediates detach to form H_2 through either the electrochemical interaction between H* and H_2O molecule with the assistance of one electron (Heyrovsky step, $H_2O + e^- + H^* \rightarrow H_2 + OH^-$) or directly by chemical recombination of two H* (Tafel step, $H^* + H^* \rightarrow H_2$) [38,39]. Based on the Tafel slope, the HER pathway could be predicted. When the Volmer step, i.e., water adsorption and dissociation, is the rate-determining step, the theoretical Tafel slope is around 120 mV/dec. If the Heyrovsky step is dominant, a Tafel slope of around 40 mV/dec is expected. When the Tafel slope is as low as 30 mV/dec, the HER predominately occurs through the Tafel step [38]. In this study, the Tafel slopes for all three samples range between 40 and 120 mV/dec. However, the value for the NF electrode is close to 120 mV/dec, indicating that the HER on NF in 1 M KOH is governed by water dissociation on the surface. In contrast, the Ni_3S_2 catalyst exhibits a Tafel slope of around 103.7 mV/dev, which suggests much faster reaction kinetics with the water dissociation reaction. Given the smallest Tafel slope (84.6 mV/dec) for the P-Ni_3S_2 catalyst, it can be apparently concluded that phosphorus doping in Ni_3S_2 has significantly accelerated the alkaline HER kinetics and rate. According to the Tafel equation, $\eta = a + b \log j$, where b is the Tafel slope, j is the current density, the exchange current density j_0, therefore, could be obtained from the Tafel slope intersection with the abscissa [17,38]. The P-Ni_3S_2 catalyst displays an exchange current density j_0 of 0.31 mA/cm^2, nearly double that of the Ni_3S_2 catalyst (0.16 mA/cm^2) and 7 times higher than that of the NF electrode (4.4×10^{-2} mA/cm^2), respectively. This further highlights the superior HER kinetics on the P-Ni_3S_2 electrode. To obtain more precise quantitative kinetic information and understand the origin of HER enhancement in P-Ni_3S_2 catalyst, electrochemical impedance spectroscopy (EIS) studies were acquired, as shown in Figure 5c. The EIS experimental data were approximated with a two CPE model equivalent circuit, from which the charge-transfer resistance R_{ct} of the HER could be extracted. The P-Ni_3S_2 catalyst demonstrates a smaller charge-transfer resistance of 5.35 Ω for the HER reaction, apparently lower than that of the Ni_3S_2 catalyst (9.61 Ω). This implies that phosphorus doping facilitates the charge transfer during the HER process.

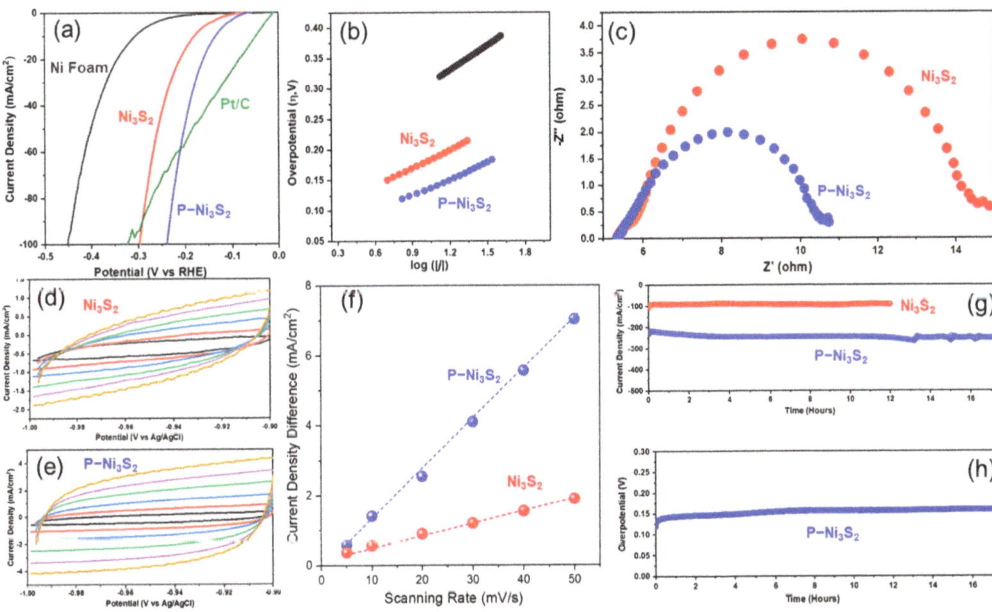

Figure 5. Electrochemical HER performances. (**a**) Polarization curves of Ni foam, Ni_3S_2 and P-Ni_3S_2, and Pt/C powder, respectively. (**b**) Tafel slopes of Ni foam, Ni_3S_2, and P-Ni_3S_2 electrodes, derived from the polarization curves in (**a**). (**c**) Impedance spectra for Ni_3S_2 and P-Ni_3S_2 at −1.3 V vs. Ag/AgCl in 1 M NaOH. (**d**,**e**) Cyclic voltammetry curves (−1~0.9 V vs. Ag/AgCl) for Ni_3S_2 and P-Ni_3S_2 electrodes at scanning rates of 5, 10, 20, 30, 40 and 50 mV/s, corresponding to CV curves arranged from inner to outer. (**f**) Current density difference at −0.95 V vs. Ag/AgCl versus the scanning rates, derived from (**d**,**e**). (**g**) Long-term stability of Ni_3S_2 and P-Ni_3S_2 electrodes under a constant applied potential of 1.5 V vs. Ag/AgCl. (**h**) Stability of P-Ni_3S_2 electrode under a constant current density of 10 mA/cm^2.

To elucidate the intrinsic origin of the HER activity improvement, the electrochemically active surface area was determined by measuring the double-layer capacitance (C_{dl}) and calculating the roughness factor R_f. The CVs were recorded within the potential range from −1.0 to −0.9 V vs. Ag/AgCl, with a scan rate varying from 5 to 50 mV/s. In this potential range, no faradaic reactions occur, and the current arises solely from the charging and discharging of the electrical double layer [17,38]. The nearly rectangular CV shape (Figure 5d,e) for both electrodes indicates that, within this potential range, the electrode works as an electrical double-layer capacitor. The difference in current density at −0.95 V vs. Ag/AgCl was plotted against the scan rate, as shown in Figure 5f, revealing a linear relationship. As a result, the C_{dl} was determined to be 34 mF/cm^2 for Ni_3S_2 and 143 mF/cm^2 for P-Ni_3S_2. Consequently, the corresponding surface roughness R_f was calculated to be 566 and 2383, respectively. Comparison of the C_{dl} and R_f reveals that the electrochemically active surface area of the P-Ni_3S_2 electrode increases by 4.2 times relative to the Ni_3S_2 electrode. The stability of the electrode catalyst is a key parameter. Figure 5g compares the chronoamperometric curves of Ni_3S_2 and P-Ni_3S_2 at 1.3 V vs. RHE. Upon application of potential, the Ni_3S_2 catalyst quickly reached a current density of ~100 mA/cm^2 and remained stable for at least 12 h of testing. In contrast, the P-Ni_3S_2 started at ~220 mA/cm^2, slightly increased, and stabilized at 250 mA/cm^2 after 18 h of testing. The P-Ni_3S_2 catalyst was further assessed for 18 h with a constant current density of 10 mA/cm^2, as shown in the chronopotentiometric curve in Figure 5h, which shows a steady potential input of ~1.13 V vs. RHE. Overall, both sulfide catalysts demonstrated robust stability, observed both under applied potential and current generation.

The electrodes were characterized following HER testing. As illustrated in Figure 6, the surface morphology remained intact overall (Figure 6a). Close-up views of both the Ni_3S_2 (Figure 6b) and P-Ni_3S_2 (Figure 6c,d) catalysts further confirm that the primary nanowire structure was preserved after HER testing. XRD analysis of these tested samples (Figure 6e) revealed that the crystal structure remained consistent with Ni_3S_2, with no obvious new phases detected.

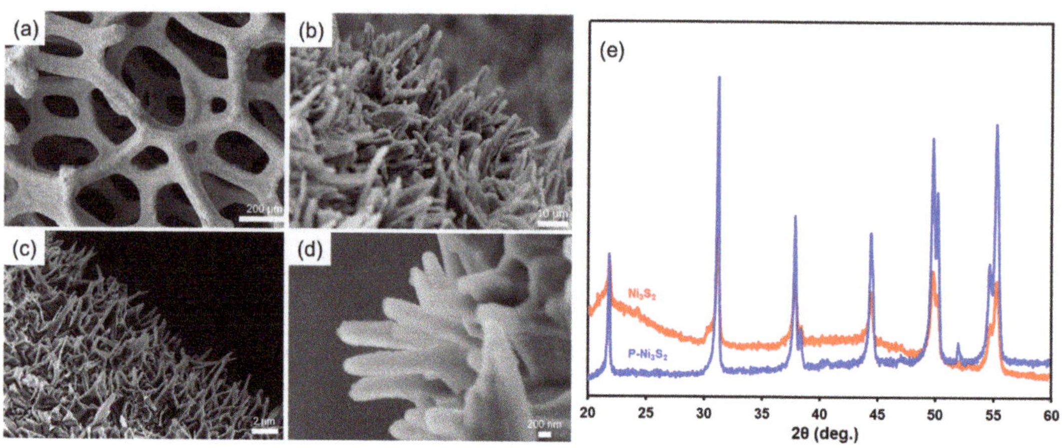

Figure 6. Surface morphology and crystal structure of Ni_3S_2 and P-Ni_3S_2 electrodes after HER testing. SEM images for Ni_3S_2 (**a**,**b**) and P-Ni_3S_2 (**c**,**d**) catalysts after HER long-term stability testing with varying magnifications. (**e**) XRD patterns for Ni_3S_2 and P-Ni_3S_2 catalysts after HER long-term stability testing.

3.4. Surface Chemistry and Electronic Structure Evolution

The surface electronic density of states (DOS) plays a crucial role in determining charge transfers and energy conversion during surface catalytic reactions [19,33]. According to *d*-band center theory, changes in the *d*-band energy levels relative to the Fermi level influence the formation of anti-bonding states, thereby affecting the adsorption strength of intermediates in electrochemical reactions. Therefore, enhancing the *d*-band center could improve the catalyst's ability to interplay with intermediates. This is typically achieved by incorporating heteroatoms to modify electronic properties [25,40]. Introducing other metals, such as Co, Mo, or Fe, into the Ni_3S_2 structure can shift the *d*-band center, optimizing hydrogen binding energy and boosting HER activity [22,24–29]. For instance, Co doping can enhance electronic conductivity and favorably adjust the *d*-band position [25,31]. To shed light on the evolution of the electronic structure, the XPS valence band spectra (VBS) for pristine Ni_3S_2 and P-Ni_3S_2 were acquired and analyzed, as shown in Figure 7. The primary peak in the valence band spectrum, located at approximately 2.3 eV, is mainly attributed to Ni 3d states, with a minor contribution from S 3p states due to hybridization between Ni 3d and S 3p orbitals. The features at 3.5 eV and 7.5 eV primarily arise from S 3p states [41]. Significant changes in the DOS near the Fermi level were observed in P-Ni_3S_2 compared to Ni_3S_2, highlighting the effect of phosphorus doping. Specifically, (i) the *d*-band shifted downward, moving further away from the Fermi level, and (ii) the DOS contribution from the *d*-band notably decreased. This trend suggests a potential degradation in the HER performance of P-Ni_3S_2, which, however, contradicts the electrochemical performance results. This discrepancy inspires us to further explore the surface chemical states and electronic structure after steady HER testing.

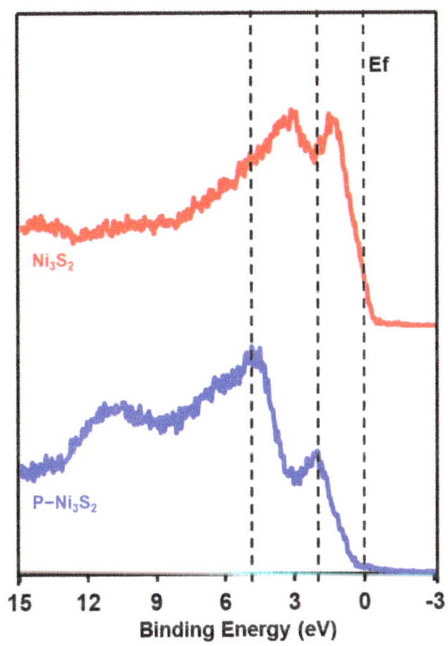

Figure 7. XPS valence band spectra (VBs) of pristine Ni_3S_2 and P-Ni_3S_2.

Surface restructuring has been observed in phosphide and chalcogenide electrocatalysts during the OER/HER process [3,6,8–13]. This phenomenon involves the in situ formation of a new layer of active species initiated on the top of the parent electrode, which plays a crucial role in governing catalytic activity during electrochemical processes. In this study, XPS depth profile characterization was employed for the first time, to the best of our knowledge, to investigate the reconstructed layer in the sulfide electrode. During depth profile acquisition, an ion gun is used to etch the material for a specific duration, after which it is turned off to allow for XPS spectra collection (Figure 8a). Each etching cycle exposes a new surface, and the XPS is used to analyze the compositions of these surfaces. Combining XPS depth profiling with ion beam etching is a powerful technique for studying the chemical composition of functional materials, spanning from the surface to the bulk [42,43]. This approach aims to provide insights into the new active species and enhance our understanding of the origins of HER improvement.

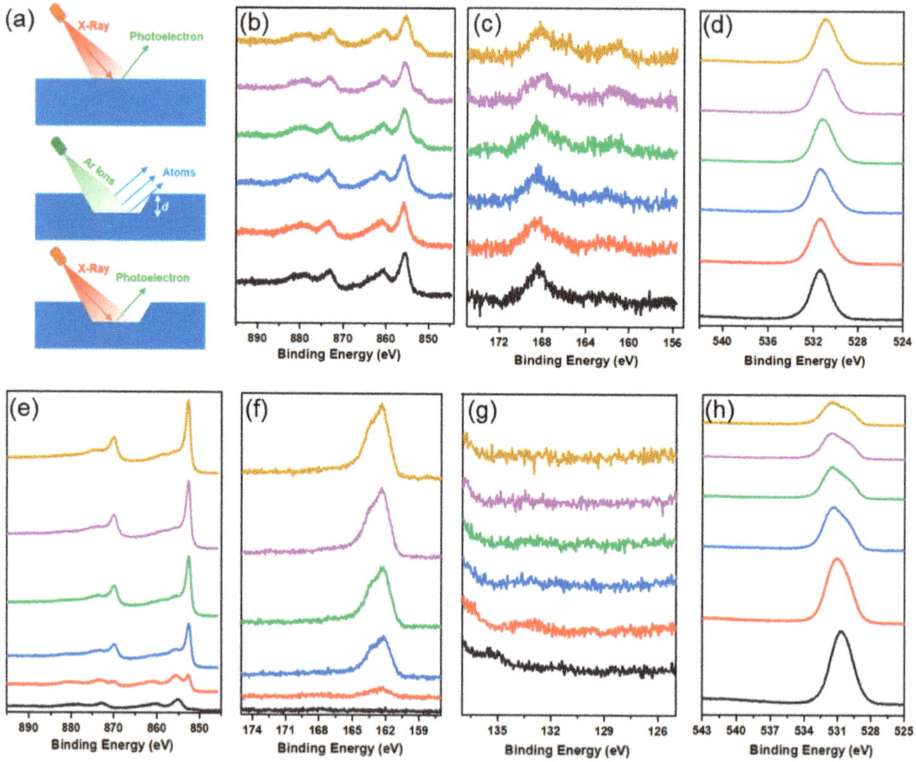

Figure 8. XPS depth profiles of Ni_3S_2 and P-Ni_3S_2 electrodes after HER long-term stability testing. (**a**) Schematic of XPS depth profiling on catalysts, high-resolution XPS depth profiling spectra for Ni 2p (**b**), S 2p (**c**), and O1s (**d**) of Ni_3S_2 catalyst; and Ni 2p (**e**), S 2p (**f**), P 2p (**g**) and O1s (**h**) of P-Ni_3S_2 catalyst, respectively. The curves, arranged from bottom to top, represent the surfaces after sputtering cycles ranging from 0 to 5.

Figure 8 presents the high-resolution XPS depth profiling spectra of the major elements in HER-tested Ni_3S_2 and P-Ni_3S_2. Notably, both electrodes exhibited distinctly different surface chemistry compared to the pristine samples. In the Ni_3S_2 electrode, Ni is primarily in a high valence state, with the major Ni 2p2/3 peak located at 855.7 eV (Figure 8b). Sulfur exhibits an S 2p binding energy of 168.4 eV (Figure 8c), which is assigned to its coordination with oxygen to form S-O bonds and is further confirmed by O 1s spectra in Figure 8d. Meanwhile, the peak for the S-Ni bond, located at around 162 eV, is very weak on the top surface and gradually becomes stronger with increasing sputtering cycles. Correspondingly, after three sputtering cycles, a subtle Ni 2p2/3 peak appears at approximately 852.1 eV, indicating the presence of the Ni-Ni bonds alongside the emergence of the S-Ni bond. For the P-Ni_3S_2 electrode, sulfur (Figure 8f) and phosphorus (Figure 8g) completely disappeared from the very top layer, while nickel predominantly existed in a high valence state, with the main Ni 2p2/3 peak located at 855.7 eV (Figure 8e). After one sputtering cycle, the S 2p signal corresponding to sulfides appeared, centered around 162 eV, while the Ni 2p2/3 peak associated with Ni-Ni bonds sharply increased and became dominant. This suggests that the layer beneath the outermost surface layer retained the

Ni_3S_2 crystal structure in the P-Ni_3S_2 electrode, as demonstrated by the XPS spectra of the pristine samples (Figure 4). Notably, the phosphorus signal did not appear even after five sputtering cycles (Figure 8g). In comparison, it is concluded that the reconstructed layer is significantly thicker in the Ni_3S_2 electrode than in the P-Ni_3S_2 electrode. This is consistent with the XRD patterns in Figure 6e, where the Ni_3S_2 electrode showed lowered diffraction intensity. Since no S and P are detected on the top surface, the presence of phosphorus likely promotes the formation of a very thin Ni-O layer, as shown in the Ni 2p (Figure 8e) and O1s (Figure 8h) spectra, which inhibits further restructuring of the inner structure and is responsible for HER activity.

The XPS depth profiles of valence band spectra were also recorded after each sputtering cycle, as shown in Figure 9. Due to surface restructuring, the VB spectra exhibited distinct differences compared to the respective pristine electrodes for both Ni_3S_2 and P-Ni_3S_2. These spectral changes indicate subtle modifications in the electronic structure [41]. The newly reconstructed layer shifts the *d*-band in the Ni_3S_2 electrode away from the Fermi level to 3.5 eV, resulting in a decrease in the density of states near the Fermi level (Figure 9a), compared to the pristine Ni_3S_2 electrode. Additionally, the VB of the Ni_3S_2 electrode remains relatively unchanged with increasing sputtering cycles. However, it is worth noting that after three sputtering cycles, a subtle shift of the major peak towards E_f occurred (Figure 9a). This observation aligns well with the emergence of Ni-Ni and S-Ni bonds shown in Figure 8b,c. In contrast, the outer Ni-O layer in the P-Ni_3S_2 electrode shows a peak at 3 eV, with a tail crossing over the Fermi level (Figure 9b), which contributes to enhanced electrical conductivity, as illustrated in the EIS spectra (Figure 5c). As revealed in Figure 9, after one sputtering cycle, the intensity of the *d*-band crossing the Fermi level significantly increased, enriching the DOS near the Fermi level. As a consequence, this optimization facilitates better binding energy of reactants and intermediates with active sites for alkaline HER, while the heterointerfaces enhance overall charge transfer within the electrode.

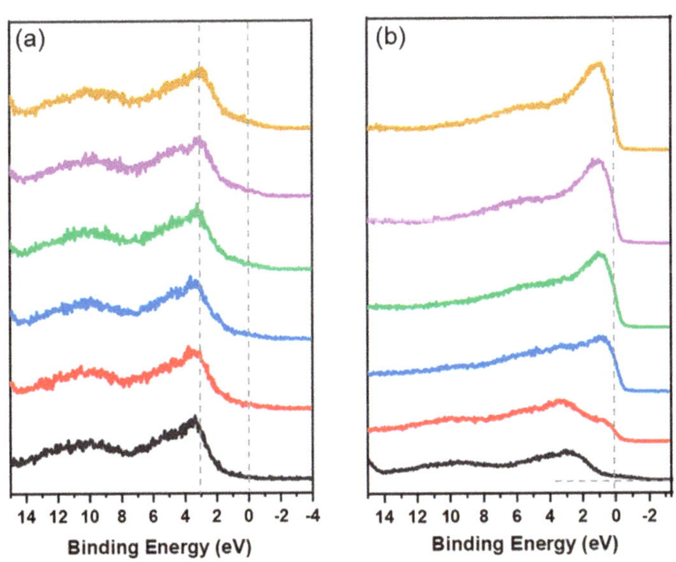

Figure 9. XPS depth profiles of valence band spectra for Ni_3S_2 (**a**) and P-Ni_3S_2 (**b**) catalysts after HER long-term stability testing. The curves, arranged from bottom to top, represent the surfaces after sputtering cycles ranging from 0 to 5.

3.5. Discussion

Surface restructuring has been confirmed on both sulfide electrodes, and it is important to highlight that the chemistry of transition-metal sulfides is highly dynamic and closely dependent on the electronic structure of the parent catalysts. In this study, phosphorus doping in Ni_3S_2 resulted in a thin reconstructed layer composed solely of Ni and O, contrasting with the pure Ni_3S_2-reconstructed active layer, which includes Ni, O, and S. This compositional difference, clearly illustrated in XPS data (Figure 8), significantly influences catalytic activity. The reconstructed Ni-O layer exhibits superior performance in the P-Ni_3S_2 electrode. Alkaline HER begins with the water dissociation process, which involves overcoming an energy barrier for H_2O rotation on the catalyst surface to achieve optimal *p-d* bonding overlap [40]. Since the Fermi level is primarily influenced by the 3*d* orbitals of Ni, the upshifted *d*-band and increased density of states near the Fermi level in the reconstructed Ni-O layer enhance its affinity for hydroxyl (OH) species, promoting water dissociation, as indicated by the smaller Tafel slope in Figure 5b. The hydrogen (H^+) generated from water dissociation will diffuse to either sulfur or oxygen sites in the Ni_3S_2-reconstructed layer, but it is more likely to adsorb on oxygen sites in the P-Ni_3S_2-reconstructed Ni-O layer. The low-coordinated oxygen sites in the Ni-O layer, formed during surface restructuring, possess strong redox capabilities for H adsorption and desorption [9,10], which contribute to the superior HER activity in the P-Ni_3S_2 electrode. Our work highlighted the crucial role of the parent catalyst in surface restructuring, emphasizing that a thorough understanding of this process is essential for designing non-precious metal sulfides to enhance catalytic activity.

4. Conclusions

In conclusion, phosphorus-doped Ni_3S_2 (P-Ni_3S_2) has been synthesized and demonstrated superior alkaline HER activity for hydrogen production. While the crystal structure and morphology were largely retained, the XPS depth profiling revealed that in situ surface restructuring occurred on electrodes during the HER process. A thick restructuring layer containing Ni, O, and S formed on the pure Ni_3S_2 electrode, whereas a very thin Ni-O layer was created on top of the P-Ni_3S_2 electrode. This in situ-formed Ni-O restructuring layer is responsible for the enhanced HER activity, featuring an increased electrochemical surface area, upshifted *d*-band, and enriched density of states near the Fermi level, which strengthen the adsorption of OH and expedite water dissociation. Furthermore, beneath the thin top Ni-O layer in the P-Ni_3S_2 electrode, the crystal structure of Ni_3S_2 was preserved, with a strong contribution from Ni 3*d* orbitals to the density of states near the Fermi level. This preservation benefits overall electrical conductivity and charge transfer by forming a heterostructure. This study highlights that a deep comprehension of surface restructuring is vital for further enhancing and predicting the catalytic activity of transition-metal sulfide electrocatalysts and calls for further investigation in this area.

Author Contributions: Conceptualization, J.L.; methodology, J.L. and D.C.; formal analysis, J.L.; investigation, J.L., C.P. and C.S.; resources, D.C.; project administration, E.A.N. and V.J.; supervision, E.A.N. and V.J.; writing—original draft preparation, J.L.; writing—review and editing, J.L., D.C., E.A.N. and V.J. All authors have read and agreed to the published version of the manuscript.

Funding: This research received no external funding.

Institutional Review Board Statement: Not applicable.

Informed Consent Statement: Not applicable.

Data Availability Statement: The original contributions presented in this study are included in the article. Further inquiries can be directed to the corresponding authors.

Acknowledgments: The authors thank support from USMA's Academic Individual Advanced Development (AIAD) program.

Conflicts of Interest: The authors declare no conflicts of interest.

References

1. Perović, I.; Kaninski, M.M.; Tasić, G.; Maslovara, S.; Laušević, P.; Seović, M.; Nikolić, V. Enhanced Catalytic Activity and Energy Savings with Ni-Zn-Mo Ionic Activators for Hydrogen Evolution in Alkaline Electrolysis. *Materials* **2023**, *16*, 5268. [CrossRef]
2. Kuang, J.; Li, Z.; Li, W.; Chen, C.; La, M.; Hao, Y. Achieving High Activity and Long-Term Stability towards Oxygen Evolution in Acid by Phase Coupling between CeO_2-Ir. *Materials* **2023**, *16*, 7000. [CrossRef] [PubMed]
3. Li, J. Oxygen Evolution Reaction in Energy Conversion and Storage: Design Strategies Under and Beyond the Energy Scaling Relationship. *Nano-Micro Lett.* **2022**, *14*, 112. [CrossRef] [PubMed]
4. Gao, L.; Cui, X.; Sewell, C.D.; Li, J.; Lin, Z. Recent advances in activating surface reconstruction for the high-efficiency oxygen evolution reaction. *Chem. Soc. Rev.* **2021**, *50*, 8428–8469. [CrossRef] [PubMed]
5. Wang, L.; Meng, Q.; Xiao, M.; Liu, C.; Xing, W.; Zhu, J. Insights into the Dynamic Surface Reconstruction of Electrocatalysts in Oxygen Evolution Reaction. *Renewables* **2024**, *2*, 272–296. [CrossRef]
6. Li, J.; Chu, D.; Baker, D.R.; Leff, A.; Zheng, P.; Jiang, R. Earth-Abundant Fe and Ni Dually Doped Co_2P for Superior Oxygen Evolution Reactivity and as a Bifunctional Electrocatalyst toward Renewable Energy-Powered Overall Alkaline Water Splitting. *ACS Appl. Energy Mater.* **2021**, *4*, 9969–9981. [CrossRef]
7. Zhu, J.; Hu, L.; Zhao, P.; Lee, L.; Wong, K. Recent Advances in Electrocatalytic Hydrogen Evolution Using Nanoparticles. *Chem. Rev.* **2020**, *120*, 851–918. [CrossRef]
8. Yan, B.; Krishnamurthy, D.; Hendon, C.H.; Deshpande, S.; Surendranath, Y.; Viswanathan, V. Surface Restructuring of Nickel Sulfide Generates Optimally Coordinated Active Sites for Oxygen Reduction Catalysis. *Joule* **2017**, *1*, 600–612. [CrossRef]
9. Li, J.; Chu, D.; Baker, D.R.; Jiang, R. Seamless separation of OH_{ad} and H_{ad} on a Ni-O catalyst toward exceptional alkaline hydrogen evolution. *J. Mater. Chem. A* **2022**, *10*, 1278. [CrossRef]
10. Ding, X.; Liu, D.; Zhao, P.; Chen, X.; Wang, H.; Oropeza, F.E.; Gorni, G.; Barawi, M.; García-Tecedor, M.; de la Peña O'Shea, V.A.; et al. Dynamic restructuring of nickel sulfides for electrocatalytic hydrogen evolution reaction. *Nat. Commun.* **2024**, *15*, 5336. [CrossRef]
11. Ma, Q.; Hu, C.; Liu, K.; Hung, S.; Ou, D.; Chen, H.; Fu, G.; Zheng, N. Identifying the electrocatalytic sites of nickel disulfide in alkaline hydrogen evolution reaction. *Nano Energy* **2017**, *41*, 148–153. [CrossRef]
12. Zhu, Y.; Chen, H.; Hsu, C.; Lin, T.; Chang, C.; Chang, S.; Tsai, L.; Chen, H. Operando unraveling of the structural and chemical stability of P-substituted $CoSe_2$ electrocatalysts toward hydrogen and oxygen evolution reactions in alkaline electrolyte. *ACS Energy Lett.* **2019**, *4*, 987–994. [CrossRef]
13. Zhai, L.; Lo, T.; Xu, Z.; Potter, J.; Mo, J.; Guo, X.; Tang, C.; Tsang, S.; Lau, S. In situ phase transformation on nickel-based selenides for enhanced hydrogen evolution reaction in alkaline medium. *ACS Energy Lett.* **2020**, *5*, 2483–2491. [CrossRef]
14. Gilbert, J.B.; Rubner, M.F.; Cohen, R.E. Depth-profiling X-ray photoelectron spectroscopy (XPS) analysis of interlayer diffusion in polyelectrolyte multilayers. *Proc. Natl. Acad. Sci. USA* **2013**, *110*, 6651–6656. [CrossRef] [PubMed]
15. Racz, A.S.; Menyhard, M. XPS depth profiling of nano-layers by a novel trial-and-error evaluation procedure. *Sci. Rep.* **2024**, *14*, 18497. [CrossRef] [PubMed]
16. Oswald, S.; Lattner, E.; Seifert, M. XPS chemical state analysis of sputter depth profiling measurements for annealed TiAl-SiO_2 and TiAl-W layer stacks. *Surf. Interface Anal.* **2020**, *52*, 924–928. [CrossRef]
17. Li, J.; Chu, D.; Dong, H.; Baker, D.R.; Jiang, R. Boosted oxygen evolution reactivity by igniting double exchange interaction in spinel oxides. *J. Am. Chem. Soc.* **2020**, *142*, 50–54. [CrossRef] [PubMed]
18. Feng, L.; Yu, G.; Wu, Y.; Li, G.; Li, H.; Sun, Y.; Asefa, T.; Chen, W.; Zou, X. High-Index Faceted Ni_3S_2 Nanosheet Arrays as Highly Active and Ultrastable Electrocatalysts for Water Splitting. *J. Am. Chem. Soc.* **2015**, *137*, 14023–14026. [CrossRef]
19. Jiang, N.; Tang, Q.; Sheng, M.; You, B.; Jiang, D.; Sun, Y. Nickel sulfides for electrocatalytic hydrogen evolution under alkaline conditions: A case study of crystalline NiS, NiS_2, and Ni_3S_2 nanoparticles. *Catal. Sci. Technol.* **2016**, *6*, 1077–1084. [CrossRef]
20. Wang, L.; Cao, J.; Lei, C.; Dai, Q.; Yang, B.; Li, Z.; Zhang, X.; Yuan, C.; Lei, L.; Hou, Y. Strongly Coupled 3D N-Doped MoO_2/Ni_3S_2 Hybrid for High Current Density Hydrogen Evolution Electrocatalysis and Biomass Upgrading. *ACS Appl. Mater. Interfaces* **2019**, *11*, 27743–27750. [CrossRef]
21. Hsieh, C.; Chuah, X.; Huang, C.; Lin, H.; Chen, Y.; Lu, S. NiFe/$(Ni,Fe)_3S_2$ core/shell nanowire arrays as outstanding catalysts for electrolytic water splitting at high current densities. *Small Methods* **2019**, *3*, 1900234. [CrossRef]
22. Liu, H.; Liu, Z.; Wang, F.; Feng, L. Efficient catalysis of N doped NiS/NiS_2 heterogeneous structure. *Chem. Eng. J.* **2020**, *397*, 125507. [CrossRef]

23. Sha, L.; Liu, T.; Ye, K.; Zhu, K.; Yan, J.; Yin, J.; Wang, G.; Cao, D. A heterogeneous interface on NiS@Ni_3S_2/$NiMoO_4$ heterostructures for efficient urea electrolysis. *J. Mater. Chem. A* **2020**, *8*, 18055–18063. [CrossRef]
24. Fang, K.; Wu, T.; Hou, B.; Lin, H. Green synthesis of Ni_3S_2 nanoparticles from a nontoxic sulfur source for urea electrolysis with high catalytic activity. *Electrochim. Acta* **2022**, *421*, 140511. [CrossRef]
25. Ying, J.; He, Z.; Chen, J.; Xiao, Y.; Yang, X. Elevating the d-Band Center of Ni_3S_2 Nanosheets by Fe Incorporation to Boost the Oxygen Evolution Reaction. *Langmuir* **2023**, *39*, 5375–5383. [CrossRef] [PubMed]
26. Yang, Y.; Zhang, K.; Lin, H.; Li, X.; Chan, H.C.; Yang, L.; Gao, Q. MoS_2-Ni_3S_2 Heteronanorods as Efficient and Stable Bifunctional Electrocatalysts for Overall Water Splitting. *ACS Catal.* **2017**, *7*, 2357–2366. [CrossRef]
27. Qin, H.; Zhang, B.; Pan, Y.; Wang, X.; Diao, L.; Chen, J.; Wu, J.; Liu, E.; Sha, J.; Ma, L.; et al. Accelerating water dissociation kinetics on Ni_3S_2 nanosheets by P-induced electronic modulation. *J. Catal.* **2020**, *381*, 493–500. [CrossRef]
28. Wu, C.; Liu, B.; Wang, J.; Su, Y.; Yan, H.; Ng, C.; Li, C.; Wei, J. 3D structured Mo-doped Ni_3S_2 nanosheets as efficient dual-electrocatalyst for overall water splitting. *Appl. Surf. Sci.* **2018**, *441*, 1024–1033. [CrossRef]
29. Lu, S.Y.; Li, S.; Jin, M.; Gao, J.; Zhang, Y. Greatly boosting electrochemical hydrogen evolution reaction over Ni_3S_2 nanosheets rationally decorated by $Ni_3Sn_2S_2$ quantum dots. *Appl. Catal. B* **2020**, *267*, 118675. [CrossRef]
30. Zheng, X.; Zhang, L.; Huang, J.; Peng, L.; Deng, M.; Li, L.; Li, J.; Chen, H.; Wei, Z. Boosting Hydrogen Evolution Reaction of Nickel Sulfides by Introducing Nonmetallic Dopants. *J. Phys. Chem. C* **2020**, *124*, 24223–24231. [CrossRef]
31. Tong, X.; Li, Y.; Pang, N.; Qu, Y.; Yan, C.; Xiong, D.; Xu, S.; Wang, L.; Chu, P.K. Co-doped Ni_3S_2 porous nanocones as high-performance bifunctional electrocatalysts in water splitting. *Chem. Eng. J.* **2021**, *425*, 130455. [CrossRef]
32. Fu, W.; Zhao, Y.; Mei, J.; Wang, F.; Han, W.; Wang, F.; Xie, E. Honeycomb-like Ni_3S_2 nanosheet arrays for high-performance hybrid supercapacitors. *Electrochim. Acta* **2018**, *283*, 737–743. [CrossRef]
33. Zhou, W.; Zhao, J.; Guan, J.; Wu, M.; Li, G. Ni_3S_2 in Situ Grown on Ni Foam Coupled with Nitrogen-Doped Carbon Nanotubes as an Efficient Electrocatalyst for the Hydrogen Evolution Reaction in Alkaline Solution. *ACS Omega* **2019**, *4*, 20244–20251. [CrossRef]
34. Chen, M.; Su, Q.; Kitiphatpiboon, N.; Zhang, J.; Feng, C.; Li, S.; Zhao, Q.; Abudula, A.; Ma, Y.; Guan, G. Heterojunction engineering of Ni_3S_2/NiS nanowire for electrochemical hydrogen evolution. *Fuel* **2023**, *331*, 125794. [CrossRef]
35. Zhou, W.; Wu, X.; Cao, X.; Huang, X.; Tan, C.; Tian, J.; Liu, H.; Wang, J.; Zhang, H. Ni_3S_2 nanorods/Ni foam composite electrode with low overpotential for electrocatalytic oxygen evolution. *Energy Environ. Sci.* **2013**, *6*, 2921–2924. [CrossRef]
36. Kang, J.S.; Park, M.; Kim, J.; Park, S.H.; Chung, D.Y.; Yu, S.; Kim, J.; Park, J.; Choi, J.; Lee, K.J.; et al. Reactively sputtered nickel nitride as electrocatalytic counter electrode for dye- and quantum dot-sensitized solar cells. *Sci. Rep.* **2015**, *5*, 10450.
37. He, W.D.; Wang, C.G.; Li, H.Q.; Deng, X.L.; Xu, X.J.; Zhai, T.Y. Ultrathin and Porous Ni_3S_2/$CoNi_2S_4$ 3D-Network Structure for Superhigh Energy Density Asymmetric Supercapacitors. *Adv. Energy Mater.* **2017**, *7*, 1700983. [CrossRef]
38. Li, J.; Chu, D.; Baker, D.R.; Dong, H.; Jiang, R.; Tran, D.T. Distorted Inverse Spinel Nickel Cobaltite Grown on a MoS_2 Plate for Significantly Improved Water Splitting Activity. *Chem. Mater.* **2019**, *31*, 7590–7600. [CrossRef]
39. Stamenkovic, V.R.; Strmcnik, D.; Lopes, P.P.; Markovic, N.M. Energy and fuels from electrochemical interfaces. *Nat. Mater.* **2017**, *16*, 57–69. [CrossRef] [PubMed]
40. An, Y.; Huang, B.; Wang, Z.; Long, X.; Qiu, Y.; Hu, J.; Zhou, D.; Lina, H.; Yang, S. Constructing three-dimensional porous Ni/Ni_3S_2 nano-interfaces for hydrogen evolution electrocatalysis under alkaline conditions. *Dalton Trans.* **2017**, *46*, 10700–10706. [CrossRef] [PubMed]
41. Krishnakumar, S.R.; Sarma, D.D. X-ray photoemission study of $NiS_{2-x}Se_x$ (x = 0.0–1.2). *Phys. Rev. B* **2003**, *68*, 155110. [CrossRef]
42. Li, D.; Chen, Y.; Zhou, C.; Shi, C.; Xu, Z.; Miao, Z.; Xi, Z.; Han, J. XPS depth profiling of functional materials: Applications of ion beam etching techniques. *Mater. Chem. Front.* **2024**, *8*, 715–731. [CrossRef]
43. Ghods, P.; Isgor, O.B.; Brown, J.R.; Bensebaa, F.; Kingston, D. XPS depth profiling study on the passive oxide film of carbon steel in saturated calcium hydroxide solution and the effect of chloride on the film properties. *Appl. Surface Sci.* **2011**, *257*, 4669–4677. [CrossRef]

materials

MDPI

Article

Transition-Metal-Doped Nickel–Cobalt Layered Double Hydroxide Catalysts for an Efficient Oxygen Evolution Reaction

Zhihan Li [1], Wenjing Yi [2], Qingqing Pang [2], Meng Zhang [1,*] and Zhongyi Liu [1,3,*]

1 College of Chemistry, Zhengzhou University, Zhengzhou 450001, China
2 School of Chemical Engineering, Zhengzhou University, Zhengzhou 450001, China
3 State Key Laboratory of Coking Coal Resources Green Exploitation, Zhengzhou University, Zhengzhou 450001, China
* Correspondence: zhangmeng1991@zzu.edu.cn (M.Z.); liuzhongyi@zzu.edu.cn (Z.L.)

Abstract: Hydrogen plays a vital role in the global shift toward cleaner energy solutions, with water electrolysis standing out as one of the most promising techniques for generating hydrogen. Despite its potential, the oxygen evolution reaction (OER) involved in this process faces significant challenges, including high overpotentials and slow reaction rates, which underscore the need for advanced electrocatalytic materials to enhance efficiency. Noble metal catalysts are effective but expensive, so transition-metal-based electrocatalysts like nickel–cobalt layered double hydroxides (NiCo LDHs) have become promising alternatives. In this research, a series of NiCo LDH catalysts doped with Fe, Mn, Cu, and Zn were effectively produced using a one-step hydrothermal technique. Among the catalysts, the Fe-doped NiCo LDH exhibited OER activity, achieving a lower overpotential (289 mV) at a current density of 50 mA/cm^2, which was far better than the 450 mV of the undoped NiCo LDH. The Mn-, Cu-, and Zn-NiCo LDHs also exhibited lower overpotentials of 414 mV, 403 mV, and 357 mV, respectively, at this current density. The Fe-doped NiCo LDH had a 3D layered nanoflower structure, increasing the surface area for reactant adsorption. The electrochemically active surface area (ECSA), as indicated by the double-layer capacitance (C_{dl}), was larger in the doped samples. The C_{dl} value of the Fe-doped NiCo LDH was 3.72 mF/cm^2, significantly surpassing the 0.82 mF/cm^2 of the undoped NiCo LDH. These changes improved charge transfer and optimized reaction kinetics, enhancing the overall OER performance. This study offers significant contributions to the development of efficient electrocatalysts for the OER, advancing the understanding of key design principles for enhanced catalytic performance.

Keywords: layered double hydroxide; transition metal; doping; hydrogen; oxygen evolution reaction

1. Introduction

Hydrogen has been increasingly recognized as a vital element in the transition to sustainable energy systems [1]. One of the primary advantages of hydrogen is its capacity to be produced from a variety of sources, including renewable energy through water electrolysis [2,3]. This method is particularly appealing since it generates hydrogen without carbon emissions, aligning with the goal of global climate change [4].

The oxygen evolution reaction (OER) holds a pivotal position among the various electrochemical processes. It is a crucial half-reaction occurring specifically at the anode during the intricate process of water electrolysis. This reaction is characterized by a significant energy barrier and sluggish kinetics, making it the rate-limiting step in the water splitting process [5,6]. The efficiency of the OER is significantly influenced by the choice of electrocatalysts, which speed the reaction by lowering the activation energy required for the process [7]. Noble metal catalysts play a main role in electrocatalysis due to their exceptional catalytic properties [8,9]. Their significant reactivity, stability, and capacity to accelerate electron transfer render them essential for advancing effective energy conversion and storage technologies [10]. However, the high cost and shortage of noble metals pose significant challenges for large-scale commercialization.

Transition metal electrocatalysts can play an alternative role in various reactions, particularly in energy conversion processes such as the OER [11,12]. These materials are favored for their unique properties, including high catalytic activity, stability, and the ability to be synthesized from abundant and inexpensive elements [13]. Among these materials, transition-metal-layered double hydroxides (TM-LDHs) stand out as particularly efficient electrocatalysts for the oxygen evolution reaction (OER), especially the (Ni, Co, Fe)-LDHs. This efficiency largely stems from their distinctive layered structures and substantial specific surface area, which provide an increased number of accessible active sites, enhance anion exchange abilities, and facilitate improved charge and mass transport characteristics. Such properties render LDHs immensely useful across various applications [14–16].

The electrolysis process enables water molecules or hydroxyl ions to be incorporated into the layered structure of layered double hydroxides (LDHs). This incorporation enhances the proton-coupled electron transfer mechanism, which is essential for enhancing the overall catalytic efficiency of LDHs. As a result, the interaction between these molecules and the layered structure not only aids in the structural integrity of the materials but also significantly boosts their efficiency in catalytic applications [17,18]. Additionally, nickel–cobalt layered double hydroxides (NiCo LDHs) have been treated as effective electrocatalysts during the OER, especially in alkaline environments [19,20]. The combination of nickel and cobalt in LDHs leads to effects that enhance their electrocatalytic properties. The combination of both metals creates an enhanced electronic environment that can improve the binding energies of reaction intermediates and optimize the OER process [21,22]. This synergy is particularly beneficial to the formation of active high-valence states during the reaction. NiCo LDHs exhibit numerous advantages; however, there still remains significant potential for enhancing their performance.

Using hydrothermal methods to synthesize NiCo LDHs can achieve better crystallinity and more uniform particle size, which are beneficial for catalytic activity. This particular method offers a remarkable advantage in that it enables meticulous and exacting control over both the composition and morphology of LDHs. When it comes to composition, it allows researchers to fine-tune the elemental ratios within the LDH structure [23]. The morphology design also affects the performance at different current densities. For instance, the hierarchical structure, such as using nickel foam as a base material, can enhance the overall conductivity of the electrode, achieving better charge transfer during the OER process. The self-supported structure not only improves its stability but also allows it to increase the active surfaces that expose more active sites to participate in the reaction [24,25]. Meanwhile, many studies have demonstrated that metal-doped NiCo LDH catalysts can lower overpotentials and accelerate OER kinetics compared to their undoped counterparts [26,27].

Transition metal elements possess abundant electron orbits and exhibit variable oxidation states [28]. This characteristic significantly influences their electronic structure when doped with NiCo LDH. When transition metal atoms are incorporated into the lattice of NiCo LDH, the differences in electronegativity and electron orbital energy levels between these atoms and the Ni and Co atoms lead to a readjustment of the electron cloud distribution [29,30]. This adjustment notably changes the electron cloud density near the active site. The electron cloud is key in chemical reactions as it probabilistically describes electron positions in atoms or molecules, and its distribution around the active site (where reactions occur) is altered upon this adjustment. For example, the nanocages Fe-Co-Ni-LDH need a significantly lower overpotential to reach a specific current density compared to other LDHs, indicating superior catalytic efficiency [31]. When a transition metal is doped, its d-orbital electrons interact with the orbitals of the reactant molecules to establish a specific adsorption mode [32]. If the distribution of electron cloud density is optimized, the reactant molecules can be adsorbed more effectively on the active sites, and the intermediate products generated during the reaction can also be converted more efficiently [33]. This optimization reduces the activation energy of the reaction and significantly enhances the rate of the catalytic reaction.

In this work, we focused on the synthesis of NiCo LDH on nickel foam through a straightforward one-step hydrothermal process. To enhance the catalytic performance, we strategically incorporated other transition metals into the NiCo LDH structure. A series of catalysts with doping elements of Fe, Mn, Cu, and Zn were successfully prepared. Notably, the Fe-NiCo LDH exhibits an impressive overpotential of 289 mV when assessed at 50 mA/cm^2. This highlights the importance of our doping modification approach in improving the electrocatalytic characteristics of NiCo LDH. The organization of this paper is as follows: Section 1 introduces the background and significance of this study. Section 2 details the experimental methods and materials. Section 3 presents the results and discussion. Finally, Section 4 concludes this study and provides future research directions.

2. Experiment

2.1. *Preparation of the Doped NiCo LDH Catalysts*

Before the experiment, the nickel foam (NF) was cleaned with ethanol, 2 mol/L hydrochloric acid, and water via ultrasonic for 15 min and dried in a vacuum at 60 °C. Then, the $Ni(NO_3)_2$ (Aladdin, China), $Co(NO_3)_2$ (Aladdin, China), and a certain amount of $FeCl_3$ (Aladdin, China) were dissolved in 30 mL of deionized water. Subsequently, 2 mmol of NH_4F and 0.5 mmol of urea were added and stirred for 10 min. Finally, the mixture was placed, along with a piece of treated foam nickel (2.5 × 3 cm), into a 50 mL Teflon-lined hydrothermal autoclave and hydrothermally treated at 120 °C for 10 h. Once the reaction finished, the autoclave was cooled to room temperature naturally. The resulting sample underwent the thorough washes with ethanol and deionized water, followed by drying at 60 °C to yield Fe-NiCo LDH. To prepare Mn-NiCo LDH, Cu-NiCo LDH, and Zn-NiCo LDH, $FeCl_3$ was replaced by a certain amount of $MnCl_2$, $CuCl_2$, and $ZnCl_2$, respectively.

2.2. *Structure Characterization*

The crystalline structures of the synthesized products were investigated by X-ray diffraction (XRD, Empyrean, PANalytical, Almelo, The Netherlands Cu Kα). The surface chemical states and electronic structures of the prepared products were probed by X-ray photoelectron spectroscopy (XPS, AXIS SUPRA, Shimadzu (Kratos Analytical), Manchester, UK). The morphologies and microstructures of the products were examined by a Scanning Electron Microscope (SEM, ZEISS Sigma 500, Carl Zeiss, Oberkochen, Germany).

2.3. *Electrochemical Performance Testing*

The electrochemical characteristics of the samples were evaluated with an electrochemical workstation (CHI 660e). Electrochemical measurements were performed in a 1.0 M KOH aqueous electrolyte using a conventional three-electrode setup. A carbon electrode functioned as the counter electrode, an Hg/HgO electrode was used as the reference electrode, and the working electrode comprised M-NiCo LDH/NF (M=Fe, Mn, Cu, Zn). Linear sweep voltammetry (LSV) and cyclic voltammetry (CV) were performed. The

Nernst equation states that $E_{RHE} = E_{Hg/HgO} + (0.098\ V + 0.059\ V \times pH)\ V$; all potentials could be converted into the reversible hydrogen electrodes (RHEs). The overpotential was calculated using the following equation $\eta = E_{RHE} - 1.23V$.

Linear sweep voltammetry (LSV) measurements were performed at a scan rate of $5\ mV\ s^{-1}$ to obtain polarization curves. The Tafel slope was calculated by making use of the equation $\eta = b \log j + a$, where η stands for the overpotential, b refers to the Tafel slope, a is the Tafel intercept, and j is equivalent to the current density ($mA\ cm^{-2}$). The electrochemical double-layer capacitance (C_{dl}) was derived from cyclic voltammetry (CV) curves within a non-Faradaic potential, calculated as $C_{dl} = \Delta j/\Delta v$, where Δv refers to the scan rate from 20 to 120 $mV\ s^{-1}$ and Δj is defined as half the difference between the anodic and cathodic current densities ($\Delta j = (j_a - j_c)/2$).

3. Results and Discussion

The SEM images depicted in Figure 1a–j reveal the morphological characteristics of the NiCo LDH and its doped variants, including the Fe-NiCo LDH, Mn-NiCo LDH, Cu-NiCo LDH, and Zn-NiCo LDH. These images clearly demonstrate the profound influence of doping on the structural morphology of NiCo LDH. In addition, the Fe-NiCo LDH and Zn-NiCo LDH display distinct morphological features that set them apart from the other samples. The Fe-NiCo LDH features a distinctive 3D layered nanoflower framework (Figure 1c,d). This distinctive structural configuration provides an extensive surface area, which enhances the adsorption capacity for reactant molecules. The hierarchical arrangement of the nanoflowers enables efficient charge transfer during the OER process [34]. The numerous exposed edges and surfaces in this structure serve as active sites, promoting the reaction. The Zn-NiCo LDH presents a 3D thorn-like snowflake structure (Figure 1i,j). The rough and non-uniform surface of the thorns increases the surface area available for the reaction. The unique morphology of the Zn-NiCo LDH contributes to its enhanced electrocatalytic performance in the OER. In contrast, the Mn-NiCo LDH and Cu-NiCo LDH (Figure 1e–h) do not exhibit such well-defined and advantageous morphologies as the Fe-NiCo LDH and Zn-NiCo LDH. This indicates that while doping positively influences the NiCo LDH, a more uniform, open, and three-dimensional morphology offers more significant advantages. These may help present new avenues for further research aimed at enhancing the active sites of catalysts.

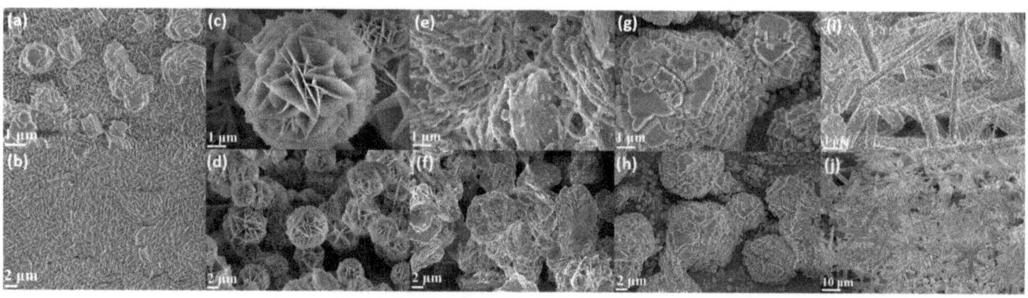

Figure 1. SEM images of (**a**,**b**) NiCo LDH, (**c**,**d**) Fe-NiCo LDH, (**e**,**f**) Mn-NiCo LDH, (**g**,**h**) Cu-NiCo LDH, and (**i**,**j**) Zn-NiCo LDH.

We characterized the composition and crystalline structure of the synthesized electrocatalysts using X-ray diffraction (XRD). The resulting patterns, as depicted in Figure 2a, closely match the reference data for nickel–cobalt layered double hydroxides (PDF#00-059-0461) and nickel foam (PDF#01-071-4654), confirming the synthesis of the target materials. In comparison to the Cu- and Zn-doped NiCo-LDHs, doping with Fe and Mn leads to more pronounced distortions in the lattice structure, resulting in notable changes to the lattice parameters. The changes in the lattice parameters can be quantitatively analyzed from the shift in the diffraction peaks. Upon doping with Fe, the peak of the Ni-Co LDH shifted to a lower angle. This observation indicates an increase in the interlayer spacing of the Ni-Co LDH structure [35]. This alteration in the lattice structure has the potential to impact the diffusion of reactant ions and the electronic structure of the material. Eventually, it will exert an influence on the catalytic performance. Specifically, lattice distortion modifies the distances between atoms and the extent of overlap within the electron cloud, consequently affecting the transmission pathways and efficiency of electrons [36]. Furthermore, lattice distortion introduces defects, which often serve as active sites that enhance the adsorption capacity of reactants and facilitate the formation and transformation of reaction intermediates, thereby improving catalytic activity.

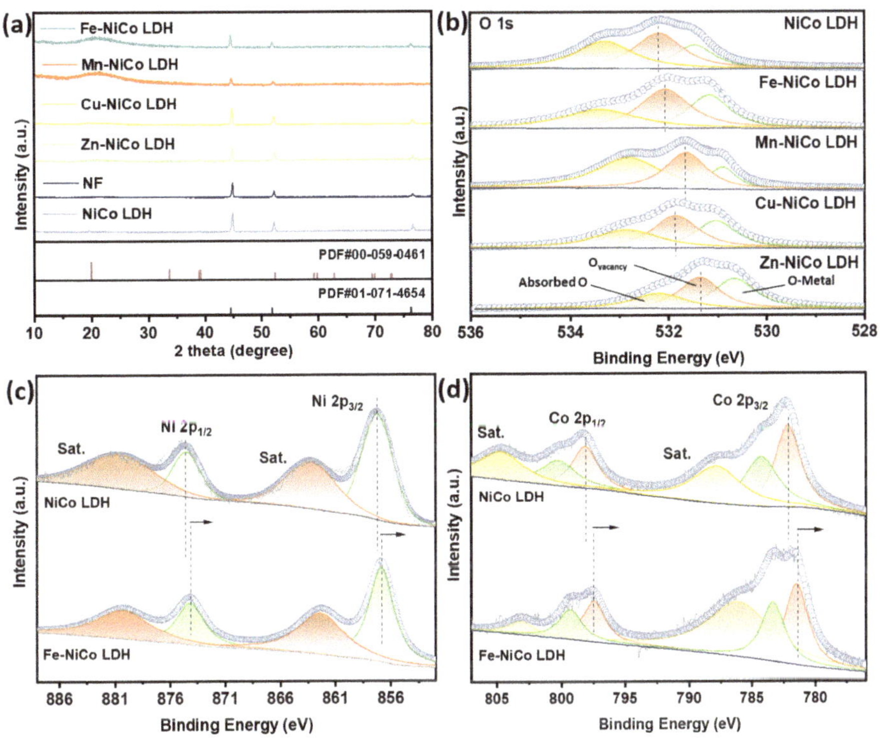

Figure 2. Structural characterization. (**a**) XRD patterns and (**b**) O1s of NiCo LDH, NF, Fe-NiCo LDH, Mn-NiCo LDH, Cu-NiCo LDH, and Zn-NiCo LDH. (**c**) Ni 2p XPS and (**d**) Co 2p XPS of Fe-NiCo LDH.

High-resolution X-ray photoelectron spectroscopy (XPS) emerged as the analytical tool that precisely determined the chemical states of elements. To guarantee the precision and dependability of the XPS data, the peak position of every element was calibrated with the aid of the C 1s peak at 284.8 eV. As depicted in Figure 2b, the high-resolution O 1s spectra acquired from all experimental samples exhibit a remarkable degree of complexity, allowing for a detailed analysis through the process of deconvolution. Specifically, these spectra can be effectively partitioned into three distinct peaks, each of which corresponds to different chemical states of oxygen present within the samples. In the case of the NiCo LDH, the peak at 533.30 eV is related to oxygen in the adsorbed state. The peak at 532.20 eV is oxygen vacancies (O_V). Meanwhile, the peak at 531.45 eV is combined with metal–oxygen bonding (M-O) [37]. These oxygen vacancies may enhance the adsorption of OH^- by providing extra adsorption sites [38]. In all the doped samples, the oxygen vacancies showed a slight negative shift compared with the NiCo LDH. This observed shift, within the context of the chemical and physical phenomena under study, implies a rather significant difference in the electronic environment that surrounds the oxygen atoms, which may be related to the interaction between the doping elements and the host NiCo LDH structure. The high-resolution Ni 2p XPS spectrum (Figure 2c) exhibits two characteristic spin–orbit split components corresponding to Ni $2p_{3/2}$ (857.1 eV) and Ni $2p_{1/2}$ (874.7 eV) states, demonstrating a peak separation of 17.6 eV, fitting with reference data for Ni^{2+} species in NiCo-LDH systems. The deconvolution of the Co 2p XPS spectrum (Figure 2d) reveals dual oxidation states, with resolved components at 784.28 eV ($2p_{3/2}$) and 800.37 eV ($2p_{1/2}$), characteristic of Co^{2+} species, contrasted by corresponding Co^{3+} signals observed at 782.18 eV ($2p_{3/2}$) and 798.14 eV ($2p_{1/2}$). This distinct separation between divalent and trivalent cobalt states aligns with established spectral fingerprints for NiCo-LDH systems. Obviously, compared with the NiCo LDH, both Ni $2p_{3/2}$ and Co $2p_{3/2}$ exhibit a negative shift not only in the Fe-NiCo LDH but in the other samples (Figure 2c,d and Figure 3a–f). The spin–orbit splitting magnitudes and principal peak locations served as diagnostic markers for identifying nickel–cobalt oxidation states while simultaneously revealing electronic configuration modifications induced by cationic substitution. The observed spectral shifts demonstrate the presence of intercomponent electronic interactions, which enable the precise modulation of the electron density distribution in the Fe-NiCo LDH and analogous heteroatom-modified systems. Such controlled electronic restructuring favors the generation of catalytically active centers while optimizing interfacial electron transport pathways between the electrode surface and oxygen evolution reaction (OER) intermediates [39]. After doping, the electron binding energies of Ni and Co changed significantly, intuitively demonstrating a modification in the electronic structure. Specifically, the d-orbital hybridization between Fe and neighboring Ni/Co atoms enhances the localized electron density at active sites [40]. This enhancement is advantageous for the adsorption of OH^- and the subsequent electron transfer process, thereby promoting the OER reaction.

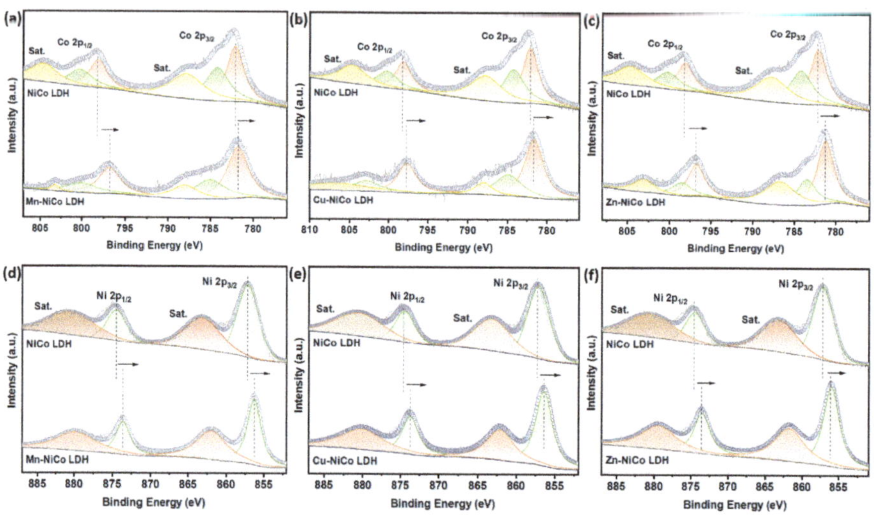

Figure 3. Co 2p XPS of (**a**) Mn-NiCo LDH, (**b**) Cu-NiCo LDH, and (**c**) Zn-NiCo LDH. Ni 2p XPS of (**d**) Mn-NiCo LDH, (**e**) Cu-NiCo LDH, and (**f**) Zn-NiCo LDH.

As shown in the Fe 2p spectra of the Fe-NiCo LDH (Figure 4a), the two peaks corresponding to Fe^{3+} are at 716.6 and 727.6 eV, while the peaks for Fe^{2+} are at 713.2 and 725.1 eV. The associated satellite peaks are located at around 720.1 and 736.9 eV. The spectral deconvolution of the Mn-NiCo LDH (Figure 4b) identifies Mn^{2+} signatures through $2p_{3/2}$ (644.3 eV) and $2p_{1/2}$ (653.3 eV) transitions in the Mn 2p region. In Cu 2p spectra of the Cu-NiCo LDH (Figure 4c), the peaks at around 932.3 and 952.1 eV are Cu $2p_{3/2}$ and Cu $2p_{1/2}$ of Cu^{2+}, respectively. The associated satellite peaks are located at around 934.9, 943.1, and 954.7 eV of strong Cu^{2+}. The Zn 2p XPS spectra of the Zn-NiCo LDH are shown in Figure 4d; the two peaks at 1021.5 and 1044.6 eV are Zn $2p_{3/2}$ and Zn $2p_{1/2}$ of Zn^{2+}, respectively. In the doped catalyst, the binding energy of Fe, Mn, Cu, and Zn (Figure 4a–d) increased compared to the general oxides. This increase may be attributed to the heightened electron cloud density surrounding the Ni and Co elements. To maintain charge balance, the electron cloud density around Fe, Mn, Cu, and Zn consequently decreased. Notably, Fe exhibits two distinct oxidation states: divalent and trivalent. The rise in binding energy contributes significantly to the stability of Fe^{2+} ions. This enhanced stability could facilitate the processes of adsorption and desorption for intermediates involved in the reactions. Furthermore, the stability of Fe^{2+} has implications for the overall reaction mechanism of the OER. A more stable Fe^{2+} can alter the pathway and kinetics of the OER, potentially leading to improved performance in catalytic applications. A more stable Fe^{2+} can act as a more efficient electron transfer mediator [41]. During the OER, the transfer of electrons between different oxidation states of Fe can improve the formation and transformation of reaction intermediates. Fe^{2+} can be oxidized to Fe^{3+} during the reaction, and then Fe^{3+} can accept electrons from the catalyst surface and be reduced back to Fe^{2+}, completing the electron transfer cycle. The enhanced stability of Fe^{2+} can ensure the smooth progress of this cycle, improving the overall catalytic performance [42].

The catalytic activity of the materials investigated for the OER was meticulously within an alkaline environment, using a 1.0 M potassium hydroxide (KOH) solution. The LSV curves for the catalysts are illustrated in Figure 5a. The Fe-NiCo LDH catalysts show remarkable electrocatalytic activity for the OER, with an overpotential of merely 289 mV at 50 mA cm^{-2}, as depicted in Figure 5b. In contrast, the undoped NiCo LDH shows higher overpotential (450 mV) at the same condition. The significant difference in overpotential clearly demonstrates a remarkable enhancement in the catalytic performance of the Fe-NiCo LDH. At a current density of 50 mA cm^{-2}, other doped samples, such as the Mn-NiCo LDH (414 mV), Cu-NiCo LDH (403 mV), and Zn-NiCo LDH (357 mV), also display lower overpotentials when compared to the undoped NiCo LDH. This clearly shows that doping positively affects the modification to the NiCo LDH for the OER.

The incorporation of Fe atoms into the NiCo LDH modifies its electronic structure, which in turn improves the capacity of active sites to capture hydroxide ions (OH^-). This improvement in adsorption capacity facilitates a more rapid reaction rate in the OER. A detailed analysis using XPS (Figure 2c,d) demonstrated that the electron binding energies for both Ni and Co registered notable changes following the doping process with Fe. This observation serves as compelling evidence of the alteration in the material's electronic structure. More specifically, the interaction between the d-orbital electrons of Fe and those of Ni and Co results in an increase in the electron cloud density surrounding the active sites, thereby promoting enhanced catalytic activity [43]. This improvement aids in the adsorption of OH^- ions and facilitates the subsequent electron transfer process, thereby significantly promoting the OER. The catalytic activity of the NiCo LDH is greatly increased through metal doping. Considering the fundamental catalytic characteristics of the NiCo LDH, the incorporation of dopants brings about diverse effects that further impact its overall performance. Coupled with the SEM analysis (Figure 1c,d), it is clear that the nanoflower structure of the Fe-NiCo LDH demonstrates the most significant performance optimization. Although the thorn-like snowflake structure of the Zn-doped NiCo LDH is different from the layered nanoflower structure of the Fe-NiCo LDH, it significantly reduces the overpotential. The surfaces of the stick exhibit roughness with non-uniformed spots (Figure 1i,j). The rough and non-uniform surface of the thorns in Zn-NiCo LDH benefits not only the surface area for OH^- adsorption but the micro-environments with different electrochemical properties. These micro-environments can improve the charge transfer process and the adsorption–desorption of reaction intermediates. From an electronic structure perspective, Zn doping may adjust the energy levels of the valence and conduction bands in the NiCo LDH. This adjustment can optimize the interaction between the catalyst and the reactants, promoting the OE. Zn doping might lower the energy barrier for the electron transfer from the catalyst to the adsorbed OH^- species, thus accelerating the oxidation process [44]. While this structure is distinctly different from the original layered configuration, it nonetheless contributes positively to performance enhancement. Although the Mn-NiCo LDH and Cu-NiCo LDH have higher overpotentials than the Fe-NiCo LDH and Zn-NiCo LDH, they still show significant improvement compared to the undoped NiCo LDH. This indicates that even though their morphologies and electronic structure modifications are not as effective as those of the Fe-NiCo LDH and Zn-NiCo LDH, the doping still positively impacts the OER activity. For the Mn-NiCo LDH, Mn doping can introduce new electronic states in the NiCo LDH. These new states can act as additional active sites or modify the existing active site properties. The variable oxidation states of Mn can participate in the redox reactions during the OER, improving the transfer of electrons [45]. However, the less-ordered morphology of the Mn-NiCo LDH might limit the accessibility of these active sites, resulting in a relatively higher overpotential when compared with the Fe- or Zn-doped NiCo LDHs. The Cu-NiCo LDH also benefits from the doping effect. Cu is known for its good electrical conductivity and ability to participate in redox reactions. When doped into the NiCo LDH, Cu can enhance the efficiency of the charge transformation within the catalyst. The Cu atoms can act as electron transport centers, increasing the movement of electrons from the bulk of the catalyst to the surface where the OER occurs.

The OER kinetics of the catalysts were evaluated by obtaining the Tafel slopes from the LSV curves (Figure 5c). The Tafel slope for the Cu-NiCo LDH was measured at 89 mV dec^{-1}, which is lower than that of the other samples, suggesting that the Cu-NiCo LDH has more rapid OER kinetics. A smaller Tafel slope implies that the reaction rate is more sensitive to the overpotential. Compared to the other catalysts, for the Cu-NiCo LDH, a relatively minor increment in the overpotential is capable of bringing about a substantial rise in the current density. Additionally, the LSV curves image (Figure 5a) shows that at 10 mA cm^{-2}, the overpotential of the Cu-NiCo LDH is the lowest. This reduction in the overpotential may be attributed to the doping of Cu, which enhances electron transport and minimizes energy loss during electron transmission [46]. Cu doping is likely to induce alterations in the reaction pathway. In the course of the OER process, reactions can progress via diverse intermediates. The existence of Cu might facilitate the generation of particular reactive intermediates. As a result, the activation energy of the reaction is decreased, and the reaction kinetics are enhanced. One possible explanation is that Cu can interact with OH^- and other reaction species to form a more reactive complex. This complex can then be more easily converted to the final product, O_2. Cu may form a Cu-OH bridge that has a lower energy barrier for the subsequent oxidation steps compared to the reaction pathway without Cu. Another point to consider is that doping with Cu can also have an impact on the surface charge distribution of the catalyst. A more favorable distribution can enhance the adsorption of OH^- and improve the desorption of reaction products, such as O_2, which can further improve the reaction kinetics.

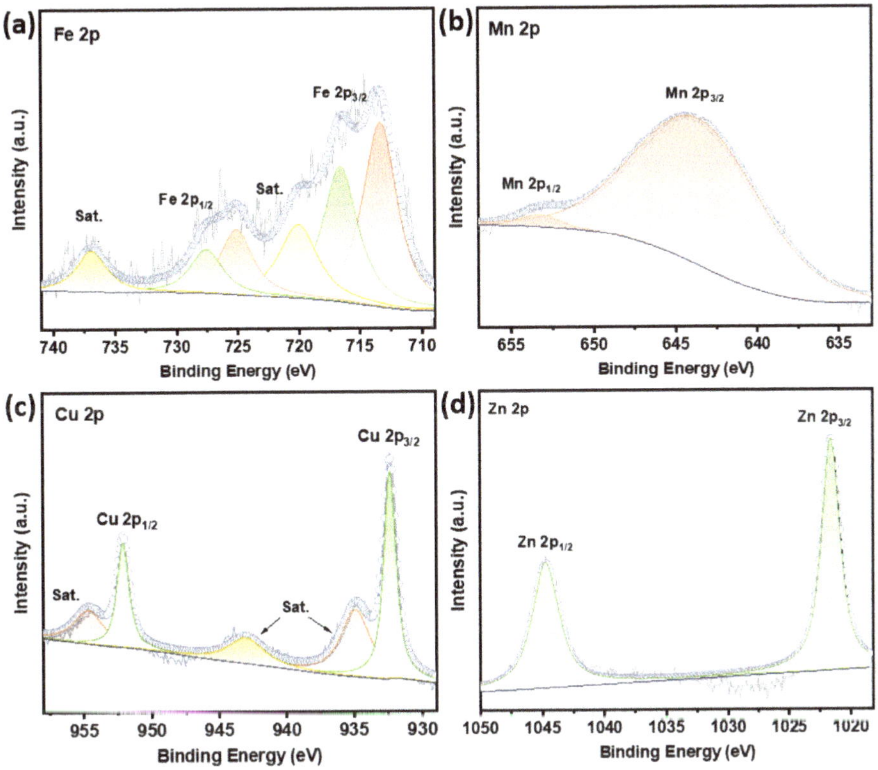

Figure 4. (**a**) Fe 2p XPS of Fe-NiCo LDH. (**b**) Mn 2p XPS of Mn-NiCo LDH. (**c**) Cu 2p XPS of Cu-NiCo LDH. (**d**) Zn 2p XPS of Zn-NiCo LDH.

In electrochemical catalysis, the double-layer capacitance (C_{dl}) is directly proportional to the electrochemically active surface area (ECSA) of the catalyst, which can be estimated by determining C_{dl} (Figure 5d). The Fe-NiCo LDH (3.72 mF cm^{-2}) has a larger C_{dl} value compared with the NiCo LDH (0.82 mF cm^{-2}), Mn-NiCo LDH (1.3 mF cm^{-2}), Cu-NiCo LDH (1.09 mF cm^{-2}), and Zn-NiCo LDH (1.05 mF cm^{-2}). The larger C_{dl} value of the Fe-NiCo LDH indicates a larger ECSA, meaning there are more exposed active sites on its surface. The unique 3D layered nanoflower structure of the Fe NiCo LDH is likely responsible for this large ECSA. The hierarchical nanoflower structure provides a high surface-to-volume ratio, allowing for a greater number of active sites to be exposed to the electrolyte. These active sites can interact with OH^- or other intermedia more effectively. The increased number of active sites leads to more available reaction sites, allowing for complete contact with electrolytes. Consequently, this promotes the reaction and enhances the reactivity of the catalyst. The enhanced performance of the Fe-doped NiCo LDH compared to the other dopants can be attributed to several key mechanisms. First, the incorporation of Fe into the NiCo LDH structure leads to significant changes in the electronic structure of the catalyst. Fe has a higher electron density and a different electronegativity compared to Ni and Co, which results in a redistribution of the electron cloud density around the active sites. This redistribution enhances the adsorption of hydroxide ions (OH^-) and facilitates the electron transfer process during the OER. Second, the Fe-doped NiCo LDH exhibits a unique 3D layered nanoflower structure, which provides a large surface area for reactant adsorption and efficient charge transfer. The hierarchical arrangement of the nanoflowers allows for better exposure of active sites and improved mass transport, which are crucial for enhancing the catalytic activity. When compared with the undoped NiCo LDH, the Mn-, Cu-, and Zn-NiCo LDHs exhibit larger C_{dl} values. This phenomenon implies that doping is able to enhance the ECSA of the NiCo LDH. Regarding the Mn-NiCo LDH, despite its morphology being less regular than that of the Fe-NiCo LDH, Mn doping can still give rise to new active sites or uncover more active sites on the surface. Mn atoms can introduce defects or modify the surface structure, leading to an increase in the ECSA. The Cu-NiCo LDH and Zn-NiCo LDH also benefit from doping in terms of the ECSA. Cu and Zn atoms can disrupt the original lattice structure of the NiCo LDH, creating more surface active sites. The unique morphology of the Zn-NiCo LDH, although different from that of the Fe-NiCo LDH, also contributes to the increase in the ECSA. The 3D thorn-like snowflake structure of the Zn-NiCo LDH provides a large surface area, which is reflected in its relatively large C_{dl} value.

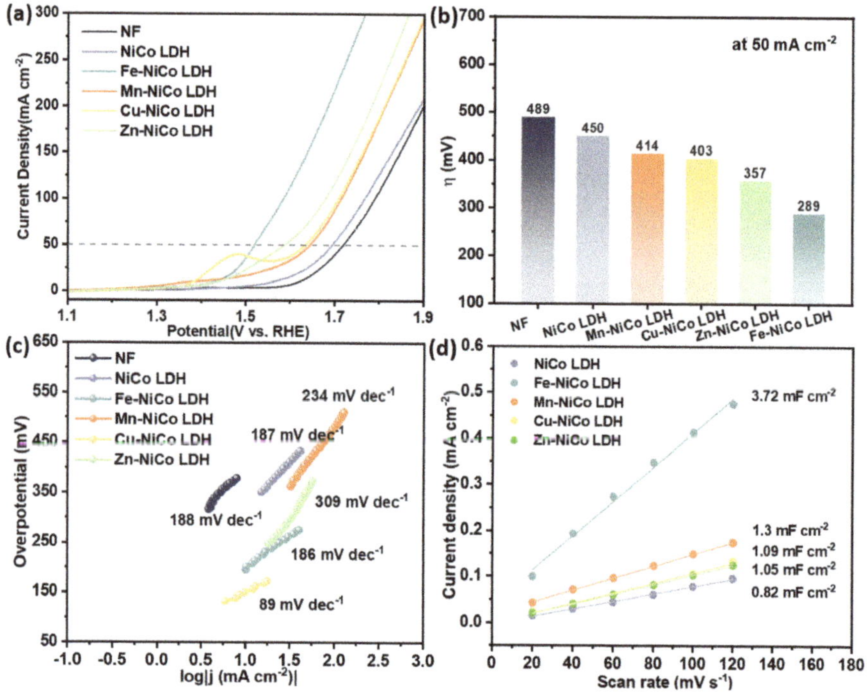

Figure 5. (**a**) LSV curves, (**b**) η at a current density of 50 mA cm^{-2}, (**c**) Tafel plots, and (**d**) linear plots of current density ($\Delta j/2$) versus scan rate for NiCo LDH, Fe-NiCo LDH, Mn-NiCo LDH, Cu-NiCo LDH, and Zn-NiCo LDH.

4. Conclusions

In this study, a series of transition-metal-doped NiCo LDH catalysts were successfully synthesized via the one-step hydrothermal method. The doping of Fe, Mn, Cu, and Zn significantly enhanced the oxygen evolution reaction (OER) performance. At a current density of 50 mA/cm^2, the undoped NiCo LDH had an overpotential of 450 mV, while the Fe-doped NiCo LDH showed a remarkable improvement, achieving an overpotential of 289 mV. The Mn-, Cu-, and Zn-NiCo LDHs also exhibited lower overpotentials of 414 mV, 403 mV, and 357 mV, respectively, at this current density. The doping modified the morphology, crystal structure, and electronic structure of the NiCo LDH. There was a unique 3D layered nanoflower in the Fe-NiCo LDH and a 3D thorn-like snowflake in the Zn-NiCo LDH, both of which increased the surface area for reactant adsorption. XRD analysis revealed that Fe and Mn doping led to lattice distortions, changing the lattice parameters. XPS revealed that doping altered the chemical states of elements and the electronic environment around them. In terms of electrochemical properties, the electrochemically active surface area, estimated by the double-layer capacitance (C_{dl}), increased in doped samples. The Fe-doped NiCo LDH had a C_{dl} value of 3.72 mF/cm^2, where the undoped NiCo LDH had a value of 0.82 mF/cm^2. The Cu-NiCo LDH exhibited the fastest OER kinetics with a Tafel slope of 89 mV dec^{-1}.

This study suggests several potential avenues for future research to further enhance OER efficiency. Future work could explore the doping of NiCo LDHs with other transition metals, such as Mo, W, or V, to identify new catalysts with even higher OER performance. The development of composite materials, such as NiCo LDHs combined with conductive polymers or carbon-based materials, could improve the electronic conductivity and stability of the catalysts. The use of advanced characterization techniques, such as in situ X-ray photoelectron spectroscopy (XPS) and X-ray absorption spectroscopy (XAS), could provide deeper insights into the electronic and geometric structures of the catalysts during the OER process. In conclusion, this study demonstrates the potential of transition-metal-doped NiCo LDH catalysts for efficient OER and provides a foundation for future research to further enhance the performance and sustainability of these catalysts.

Author Contributions: Conceptualization, M.Z. and Z.L. (Zhongyi Liu); Methodology, Z.L. (Zhihan Li), W.Y., Q.P. and M.Z.; Software, W.Y.; Formal analysis, Z.L. (Zhihan Li); Investigation, Z.L. (Zhihan Li), W.Y. and Q.P.; Resources, Z.L. (Zhongyi Liu); Data curation, W.Y.; Writing—original draft, Z.L. (Zhihan Li); Writing—review & editing, Q.P., M.Z. and Z.L. (Zhongyi Liu); Supervision, Z.L. (Zhongyi Liu); Project administration, M.Z.; Funding acquisition, Z.L. (Zhongyi Liu). All authors have read and agreed to the published version of the manuscript.

Funding: This research received no external funding.

Institutional Review Board Statement: Not applicable.

Informed Consent Statement: Not applicable.

Data Availability Statement: The original contributions presented in the study are included in the article, further inquiries can be directed to the corresponding author.

Conflicts of Interest: The authors declare no conflict of interest.

Abbreviations and Full Forms

LDH: layered double hydroxide; OER: oxygen evolution reaction; ECSA: electrochemically active surface area; C_{dl}: double-layer capacitance; SEM: Scanning Electron Microscope; XRD: X-ray diffraction; XPS: X-ray photoelectron spectroscopy; LSV: linear sweep voltammetry; Tafel: Tafel slope; RHE: reversible hydrogen electrode; NF: nickel foam; M-NiCo LDH: transition-metal-doped nickel–cobalt layered double hydroxide (where M represents Fe, Mn, Cu, or Zn).

References

1. Zhang, W.; Liu, M.; Gu, X.; Shi, Y.; Deng, Z.; Cai, N. Water electrolysis toward elevated temperature: Advances, challenges and frontiers. *Chem. Rev.* **2023**, *123*, 7119–7192. [CrossRef] [PubMed]
2. Chi, J.; Yu, H. Water electrolysis based on renewable energy for hydrogen production. *Chin. J. Catal.* **2018**, *39*, 390–394. [CrossRef]
3. Zhao, Y.; Niu, Z.; Zhao, J.; Xue, L.; Fu, X.; Long, J. Recent advancements in photoelectrochemical water splitting for hydrogen production. *Electrochem. Energy Rev.* **2023**, *6*, 14. [CrossRef]

4. Yan, Z.; Hitt, J.L.; Turner, J.A.; Mallouk, T.E. Renewable electricity storage using electrolysis. *Proc. Natl. Acad. Sci. USA* **2020**, *117*, 12558–12563. [CrossRef]
5. Lee, W.H.; Ko, Y.-J.; Kim, J.-Y.; Min, B.K.; Hwang, Y.J.; Oh, H.-S. Single-atom catalysts for the oxygen evolution reaction: Recent developments and future perspectives. *Chem. Commun.* **2020**, *56*, 12687–12697. [CrossRef]
6. Wu, H.; Huang, Q.; Shi, Y.; Chang, J.; Lu, S. Electrocatalytic water splitting: Mechanism and electrocatalyst design. *Nano Res.* **2023**, *16*, 9142–9157. [CrossRef]
7. Zhang, B.; Shan, J.; Wang, W.; Tsiakaras, P.; Li, Y. Oxygen vacancy and core–shell heterojunction engineering of anemone-like CoP@CoOOH bifunctional electrocatalyst for efficient overall water splitting. *Small* **2022**, *18*, 2106012. [CrossRef]
8. Wu, J.-X.; Chen, W.-X.; He, C.-T.; Zheng, K.; Zhuo, L.-L.; Zhao, Z.-H.; Zhang, J.-P. Atomically dispersed dual-metal sites showing unique reactivity and dynamism for electrocatalysis. *Nano-Micro Lett.* **2023**, *15*, 120. [CrossRef] [PubMed]
9. Li, P.; Wang, M.; Duan, X.; Zheng, L.; Cheng, X.; Zhang, Y.; Kuang, Y.; Li, Y.; Ma, Q.; Feng, Z.; et al. Boosting oxygen evolution of single-atomic ruthenium through electronic coupling with cobalt-iron layered double hydroxides. *Nat. Commun.* **2019**, *10*, 1711. [CrossRef]
10. Pan, Y.; Li, Y.; Nairan, A.; Khan, U.; Hu, Y.; Wu, B.; Sun, L.; Zeng, L.; Gao, J. Constructing FeNiPt@C trifunctional catalyst by high spin-induced water oxidation activity for Zn-air battery and anion exchange membrane water electrolyzer. *Adv. Sci.* **2024**, *11*, 2308205. [CrossRef]
11. Das, C.; Sinha, N.; Roy, P. Transition metal non-oxides as electrocatalysts: Advantages and challenges. *Small* **2022**, *18*, 2202033. [CrossRef] [PubMed]
12. Han, L.; Dong, S.; Wang, E. Transition-metal (Co, Ni, and Fe)-based electrocatalysts for the water oxidation reaction. *Adv. Mater.* **2016**, *28*, 9266–9291. [CrossRef]
13. Chen, D.; Pu, Z.; Lu, R.; Ji, P.; Wang, P.; Zhu, J.; Lin, C.; Li, H.; Zhou, X.; Hu, Z.; et al. Ultralow Ru loading transition metal phosphides as high-efficient bifunctional electrocatalyst for a solar-to-hydrogen generation system. *Adv. Energy Mater.* **2020**, *10*, 2000814. [CrossRef]
14. Fan, G.; Li, F.; Evans, D.G.; Duan, X. Catalytic applications of layered double hydroxides: Recent advances and perspectives. *Chem. Soc. Rev.* **2014**, *43*, 7040–7066. [CrossRef]
15. Zhang, X.; Yan, F.; Ma, X.; Zhu, C.; Wang, Y.; Xie, Y.; Chou, S.; Huang, Y.; Chen, Y. Regulation of morphology and electronic structure of FeCoNi layered double hydroxides for highly active and stable water oxidization catalysts. *Adv. Energy Mater.* **2021**, *11*, 2102141. [CrossRef]
16. Wang, Q.; O'Hare, D. Recent advances in the synthesis and application of layered double hydroxide (LDH) nanosheets. *Chem. Rev.* **2012**, *112*, 4124–4155. [CrossRef]
17. Jiang, H.; Qin, H.; Zhou, P.; Kong, L.; Wang, C.; Ji, Z.; Shen, X.; Zhu, G.; Yuan, A. Partial sulfidation strategy to NiCo-LDH@NiCoS coupled with NiFe-LDH for highly efficient overall water splitting. *Int. J. Hydrogen Energy* **2024**, *58*, 892–901. [CrossRef]
18. Han, H.; Kim, K.M.; Ryu, J.H.; Lee, H.J.; Woo, J.; Ali, G.; Chung, K.Y.; Kim, T.; Kang, S.; Choi, S.; et al. Boosting oxygen evolution reaction of transition metal layered double hydroxide by metalloid incorporation. *Nano Energy* **2020**, *75*, 104945. [CrossRef]
19. Morales-Guio, C.G.; Liardet, L.; Hu, X. Oxidatively electrodeposited thin-film transition metal (oxy)hydroxides as oxygen evolution catalysts. *J. Am. Chem. Soc.* **2016**, *138*, 8946–8957. [CrossRef] [PubMed]
20. Jiang, J.; Zhang, A.; Li, L.; Ai, L. Nickel–cobalt layered double hydroxide nanosheets as high-performance electrocatalyst for oxygen evolution reaction. *J. Power Sources* **2015**, *278*, 445–451. [CrossRef]
21. Sanchis-Gual, R.; Hunt, D.; Jaramillo-Hernández, C.; Seijas-Da Silva, A.; Mizrahi, M.; Marini, C.; Oestreicher, V.; Abellán, G. Crystallographic and geometrical dependence of water oxidation activity in Co-based layered hydroxides. *ACS Catal.* **2023**, *13*, 10351–10363. [CrossRef]
22. Wang, Z.; Goddard, W.A.; Xiao, H. Potential-dependent transition of reaction mechanisms for oxygen evolution on layered double hydroxides. *Nat. Commun.* **2023**, *14*, 4228. [CrossRef]
23. Yuan, J.; Li, Y.; Lu, G.; Gao, Z.; Wei, F.; Qi, J.; Sui, Y.; Yan, Q.; Wang, S. Controlled synthesis of flower-like hierarchical NiCo-layered double hydroxide integrated with metal–organic framework-derived Co@C for supercapacitors. *ACS Appl. Mater. Interfaces* **2023**, *15*, 36143–36153. [CrossRef]
24. Yang, Y.; Xie, Y.; Yu, Z.; Guo, S.; Yuan, M.; Yao, H.; Liang, Z.; Lu, Y.R.; Chan, T.-S.; Li, C.; et al. Self-supported NiFe-LDH@CoSx nanosheet arrays grown on nickel foam as efficient bifunctional electrocatalysts for overall water splitting. *Chem. Eng. J.* **2021**, *419*, 129512. [CrossRef]
25. Li, Y.-K.; Zhang, G.; Lu, W.-T.; Cao, F.-F. Amorphous Ni–Fe–Mo suboxides coupled with Ni network as porous nanoplate array on nickel foam: A highly efficient and durable bifunctional electrode for overall water splitting. *Adv. Sci.* **2020**, *7*, 1902034. [CrossRef]
26. Xia, X.; Wang, S.; Liu, D.; Wang, F.; Zhang, X.; Zhang, H.; Yu, X.; Pang, Z.; Li, G.; Chen, C.; et al. Electronic modulation in Cu doped NiCo LDH/NiCo heterostructure for highly efficient overall water splitting. *Small* **2024**, *20*, 2311182. [CrossRef] [PubMed]

27. Zhao, G.; Mu, L.; Chen, L.; Chen, H.; Liao, W.; Zhao, N.; Zhang, B.; Du, M.; Zhang, T. Construction of NiFe/MnNiCo-LDH heterostructure for enhanced oxygen evolution reaction for efficient water splitting. *Int. J. Hydrogen Energy* **2024**, *88*, 432–440. [CrossRef]
28. Song, F.; Bai, L.; Moysiadou, A.; Lee, S.; Hu, C.; Liardet, L.; Hu, X. Transition Metal Oxides as Electrocatalysts for the Oxygen Evolution Reaction in Alkaline Solutions: An Application-Inspired Renaissance. *J. Am. Chem. Soc.* **2018**, *140*, 7748–7759. [CrossRef] [PubMed]
29. Li, S.; Yue, C.; Wang, H.; Du, J.; Cui, H.; Pu, M.; Lei, M. Transition metal doped into layered double hydroxide as efficient electrocatalysts for oxygen evolution reaction: A DFT study. *Appl. Surf. Sci.* **2025**, *679*, 161233. [CrossRef]
30. Cao, H.; Liu, B.; Bai, J.; Li, C.; Xu, G. Interfacial engineering of hierarchical ultra-thin NiCo-LDH nanosheet superstructures nanofiber for water cracking electrocatalysis. *J. Alloys Compd.* **2025**, *1010*, 178041. [CrossRef]
31. Zhang, J.; Yu, L.; Chen, Y.; Lu, X.F.; Gao, S.; Lou, X.W. (David) Designed formation of double-shelled Ni–Fe layered-double-hydroxide nanocages for efficient oxygen evolution reaction. *Adv. Mater.* **2020**, *32*, 1906432. [CrossRef] [PubMed]
32. He, C.; Lee, C.-H.; Meng, L.; Chen, H.-Y.T.; Li, Z. Selective orbital coupling: An adsorption mechanism in single-atom catalysis. *J. Am. Chem. Soc.* **2024**, *146*, 12395–12400. [CrossRef] [PubMed]
33. Mahsud, A.; Arif, M.; Khan, W.U.; Zhang, T.; Hussain, S.; Azam, M.; Lu, Z. Theoretical investigation of d-orbitals involvement in catalytic activity by incorporation of bimetallic on adjacent position. *Mol. Catal.* **2023**, *550*, 113526. [CrossRef]
34. Zou, Y.; Liu, C.; Zhang, C.; Yuan, L.; Li, J.; Bao, T.; Wei, G.; Zou, J.; Yu, C. Epitaxial growth of metal-organic framework nanosheets into single-crystalline orthogonal arrays. *Nat. Commun.* **2023**, *14*, 5780. [CrossRef]
35. Liu, L.; Dai, J.; Das, S.; Wang, Y.; Yu, H.; Xi, S.; Zhang, Z.; Tu, X. Plasma-catalytic CO_2 reforming of toluene over hydrotalcite-derived NiFe/(Mg, Al)O_x catalysts. *JACS Au* **2023**, *3*, 785–800. [CrossRef] [PubMed]
36. Du, S.; Ren, Z.; Wang, X.; Wu, J.; Meng, H.; Fu, H. Controlled atmosphere corrosion engineering toward inhomogeneous NiFe-LDH for energetic oxygen evolution. *ACS Nano* **2022**, *16*, 7794–7803. [CrossRef] [PubMed]
37. Liang, W.; Wang, M.; Ma, C.; Wang, J.; Zhao, C.; Hong, C. NiCo-LDH hollow nanocage oxygen evolution reaction promotes luminol electrochemiluminescence. *Small* **2024**, *20*, 2306473. [CrossRef]
38. Chu, H.; Zhang, D.; Feng, P.; Gu, Y.; Chen, P.; Pan, K.; Xie, H.; Yang, M. Engineering oxygen vacancies in CoO@Co_3O_4/C nanocomposites for enhanced electrochemical performances. *Nanoscale* **2021**, *13*, 19518–19526. [CrossRef]
39. Weidler, N.; Paulus, S.; Schuch, J.; Klett, J.; Hoch, S.; Stenner, P.; Maljusch, A.; Brötz, J.; Wittich, C.; Kaiser, B.; et al. CoOx thin film deposited by CVD as efficient water oxidation catalyst: Change of oxidation state in XPS and its correlation to electrochemical activity. *Phys. Chem. Chem. Phys.* **2016**, *18*, 10708–10718. [CrossRef]
40. He, Z.-D.; Tesch, R.; Eslamibidgoli, M.J.; Eikerling, M.H.; Kowalski, P.M. Low-spin state of Fe in Fe-doped NiOOH electrocatalysts. *Nat. Commun.* **2023**, *14*, 3498. [CrossRef]
41. Bao, F.; Kemppainen, E.; Dorbandt, I.; Xi, F.; Bors, R.; Maticiuc, N.; Wenisch, R.; Bagacki, R.; Schary, C.; Michalczik, U.; et al. Host, suppressor, and promoter—The roles of Ni and Fe on oxygen evolution reaction activity and stability of NiFe alloy thin films in alkaline media. *ACS Catal.* **2021**, *11*, 10537–10552. [CrossRef]
42. Zeng, H.; Zhao, X.; Zhao, F.; Park, Y.; Sillanpää, M. Accelerated Fe^{3+}/Fe^{2+} cycle using atomic H^* on pd/Al_2O_3: A novel mechanism for an electrochemical system with particle electrode for iron sludge reduction in the Fe^{2+}/peroxydisulfate oxidation process. *Chem. Eng. J.* **2020**, *382*, 122972. [CrossRef]
43. Ou, Y.; Twight, L.P.; Samanta, B.; Liu, L.; Biswas, S.; Fehrs, J.L.; Sagui, N.A.; Villalobos, J.; Morales-Santelices, J.; Antipin, D.; et al. Cooperative Fe sites on transition metal (oxy)hydroxides drive high oxygen evolution activity in base. *Nat. Commun.* **2023**, *14*, 7688. [CrossRef]
44. Zhang, L.; Luo, Y.; Wang, Q.; Wei, D.; Hu, H.; Yan, P.; Feng, X.; Qiu, S.; Xu, F.; Cao, W.; et al. Zn doping induced local lattice expansion for hierarchical hollow NiCo-LDH toward enhanced electrochemical performance in asymmetric supercapacitors. *J. Energy Storage* **2024**, *92*, 112195. [CrossRef]
45. Zhang, Z.; Huo, H.; Yu, Z.; Xiang, L.; Xie, B.; Du, C.; Wang, J.; Yin, G. Unraveling the reaction mechanism of low dose Mn dopant in $Ni(OH)_2$ supercapacitor electrode. *J. Energy Chem.* **2021**, *61*, 497–506. [CrossRef]
46. Li, Y.-H.; Wang, Y.-F.; Li, J.-Y.; Qi, M.-Y.; Conte, M.; Tang, Z.-R.; Xu, Y.-J. Selective imines synthesis by designing an atomic-level Cu–Pt electron transfer channel over CdS nanosheets. *ACS Catal.* **2024**, *14*, 657–669. [CrossRef]

materials

MDPI

Article

Nickel Electrocatalysts Obtained by Pulsed Current Electrodeposition from Watts and Citrate Baths for Enhanced Hydrogen Evolution Reaction in Alkaline Media

Raluca Bojîncă [1], Roxana Muntean [2], Rebeca Crişan [1] and Andrea Kellenberger [1,*]

[1] Faculty of Chemical Engineering, Biotechnologies and Environmental Protection, Politehnica University Timişoara, Piata Victoriei No. 2, 300006 Timişoara, Romania; raluca.bojinca@student.upt.ro (R.B.); rebeca9803@gmail.com (R.C.)

[2] Department of Materials and Manufacturing Engineering, Politehnica University Timişoara, Piata Victoriei No. 2, 300006 Timişoara, Romania; roxana.muntean@upt.ro

* Correspondence: andrea.kellenberger@upt.ro; Tel.: +40-256-404176

Abstract: Efficient and low-cost electrocatalysts for the hydrogen evolution reaction (HER) in alkaline media are essential for sustainable hydrogen production. In this study, Ni electrocatalysts were deposited on pencil graphite using a simple one-step pulsed current electrodeposition method, from both acidic Watts and alkaline citrate baths. The influence of bath type and electrodeposition parameters—current density and temperature—on catalyst morphology and performance for HER was systematically investigated by scanning electron microscopy and electrochemical methods. Linear sweep voltammetry, chronopotentiometry, and electrochemical impedance spectroscopy (EIS) were used to evaluate the electrocatalytic activity, stability, and HER mechanism. The best catalytic performance was achieved for the Ni electrocatalyst deposited from the citrate bath at 50 mA cm^{-2} and 40 °C, showing an exchange current density of 0.93 mA cm^{-2}, a Tafel slope of −208 mV dec^{-1}, and overpotentials of −210 mV and −386 mV at 10 and 100 mA cm^{-2}, respectively, in 1 M KOH solution. Chronopotentiometry confirmed improved stability and an overpotential reduction of approximately 92 mV as compared to pure Ni, while EIS revealed the lowest charge transfer resistance. It was shown that the electrocatalysts deposited from the citrate bath outperform those from the Watts bath, and electrodeposition at 40 °C is optimal for achieving the highest electrocatalytic activity for HER.

Keywords: hydrogen evolution reaction; alkaline media; nickel electrodeposition; citrate bath; pulsed current electrodeposition

1. Introduction

In recent years, the focus on renewable energy technologies has significantly propelled research into advanced electrolysis systems, particularly on anion exchange membrane (AEM) electrolyzers, which possess substantial potential for efficient hydrogen production. The core reactions at the cathode and anode are the hydrogen evolution reaction (HER) and oxygen evolution reaction (OER), respectively, both being successfully catalyzed by the benchmark electrocatalysts based on platinum, iridium, or ruthenium [1,2]. Besides these, nickel and its alloys have emerged as promising non-precious metal electrocatalysts for HER due to their remarkable intrinsic properties and electrochemical stability under operational conditions, especially in alkaline media [3,4]. Its abundance, cost-effectiveness, and favorable catalytic activity make it an attractive alternative to noble metal catalysts [5,6], which, despite their high efficiency, are hindered by scarcity and expense. In alkaline environments, the HER process is often limited by the sluggish kinetics of water dissociation. In this regard, to further enhance the activity and performance, researchers have explored various strategies, including alloying nickel with other metals [7,8] to form compounds like sulfides [9,10], selenides [11], phosphides [12] and oxides [13,14] as well as engineering nanostructures to increase the electrochemically active surface area [15]. Feng et al. reported a nanoscale electrocatalyst composed of NiS_2 and $Ni_3C@C$, presenting exceptional stability and HER performance superior to many traditional catalysts [9]. Similarly, Liu et al. suggested nickel phosphides, which facilitate hydrogen adsorption and desorption, thereby accelerating reaction rates during HER, highlighting them as effective catalysts for alkaline electrolysis [12]. Sadhanala et al. illustrated that the Ru-doped nickel foam exhibits enhanced performance due to synergistic effects, improving stability and catalytic efficiency in alkaline environments [16]. In addition to these materials, nickel-based nitrides have attracted considerable interest due to their promising electrochemical properties. The study of Wang et al. demonstrated that nitride materials, particularly nickel–cobalt nitrides, show excellent catalytic activity for both HER and OER, underscoring their versatility in alkaline electrolysis setups [17]. Their effective performance is attributed to improved electronic properties and enhanced water dissociation kinetics, which are vital for efficient hydrogen production. One of the notable approaches to improving nickel-based catalysts lies in the development of bimetallic and heterostructured materials. For instance, the combination of nickel with transition metals such as molybdenum has shown significant enhancement in HER activity [18,19]. The study of Fernández-Valverde et al. demonstrated that Ni-Mo electrocatalysts possess advantageous electrochemical behavior in alkaline solutions, establishing them as references in commercial alkaline electrolyzers due to their superior efficiency compared to pure nickel [20]. Furthermore, incorporating heteroatoms into nickel-based structures has proven to be an effective strategy to increase overall HER performance, as discussed by Deng et al., which optimizes the electronic structure and stability of the catalysts [21]. These findings are supported by Wang et al.'s study, which showed that synergetic effects between bimetallic phosphide catalysts can significantly reduce overpotentials, thereby enhancing hydrogen production efficiency [22]. Moreover, advancements in the architecture of nickel catalysts have further improved their electrochemical performance. Li et al. reported on the efficient water splitting facilitated by three-dimensional nickel molybdenum phosphide hetero-nanorods, emphasizing that the rough surface texture provided a high density of active sites, enabling rapid electron and proton transfer [23]. Such three-dimensional structures allow for effective mass transport and increased active surface area, leading to enhanced catalytic efficiency. Additional studies indicate that variation of compositions and structures, such as nickel hydroxides or oxides, can facilitate superior catalytic properties, with amorphous metallic nickel being highlighted for its impressive HER activity with minimal overpotential [24].

Among different manufacturing techniques, electrodeposition has been employed to fabricate nickel coatings [25] or particles [26], with controlled textures and porosities, which are crucial for optimizing HER performance. Electrodeposition involves the reduction of Ni^{2+} ions from an electrolyte solution onto a conductive substrate under an applied electric potential. The process parameters—such as electrolyte composition, pH, temperature, deposition potential, and current density—play pivotal roles in determining the nucleation and growth mechanisms of Ni, thereby affecting the microstructure and catalytic properties of the deposited layer [27–29]. A promising technique for tailoring the microstructure of electrodeposited materials is pulsed current (PC) electrodeposition. By alternating the applied current between a high value during the on time (t_{on}) and zero or low current during the off time (t_{off}), this method enhances nucleation rate and offers better control over growth, leading to finer-grained deposits with improved structural uniformity and electrochemical performance compared to conventional direct current electrodeposition [30]. Optimizing pulse parameters allows obtaining nanocrystalline materials with superior catalytic activity due to their favorable morphology, increased surface area, improved charge transfer kinetics, and more active sites available for the HER [31,32].

A significant role in enhancing the activity, stability, and electrochemical properties of the catalysts is played by the support material. Carbon-based supports such as graphene, carbon nanotubes, or carbon nanofibers are commonly used due to their high electrical conductivity, large surface area, and chemical stability across various pH ranges. Among them, pencil graphite electrodes (PGE) are frequently used in sensor applications and electrocatalysis [33,34], including HER [32,35], based on their low cost, accessibility, and suitability as a scalable platform for preliminary electrochemical studies.

In this study, we report on nickel deposits obtained on PGE substrate by pulsed current electrodeposition, from acidic Watts and alkaline citrate baths, with enhanced electrocatalytic activity for HER in alkaline media. We investigated the effect of temperature and different bath compositions on the electrocatalytic activity for HER. Field emission scanning electron microscopy (FE-SEM) and energy dispersive X-ray spectroscopy (EDAX) were employed to investigate surface morphology and chemical composition. The electrocatalytic performance for HER was evaluated and compared using linear sweep voltammetry (LSV) and kinetic parameters extracted from the Tafel plots. Additionally, the overpotential required to achieve current densities of 10, 50, and 100 mA cm^{-2} served as an indicator for comparing the activity of all Ni-PGE electrocatalysts. Stability tests were performed at constant current densities of 10 and 100 mA cm^{-2}. In addition, electrochemical impedance spectroscopy (EIS) was used to gain deeper insight into the electrode behavior and underlying HER mechanism at various overpotential values. The results demonstrate that Ni electrocatalysts prepared from an alkaline citrate bath at an optimal deposition temperature of 40 °C exhibit superior performance for HER. These findings provide a foundation for the development and design of HER electrocatalysts with improved electrocatalytic activity in alkaline media, utilizing a simple, one-step electrodeposition approach and an affordable and available substrate.

2. Materials and Methods

2.1. Preparation of Ni-PGE by Pulsed Current Electrodeposition

Nickel electrodeposition on PGE was carried out via pulsed current electrodeposition, from citrate and Watts bath, respectively, with the compositions described in Table 1. For the electrodeposition, a three-electrode cell was used, with PGE as working electrode, a silver—silver chloride (Ag/AgCl in 3 M KCl, BASi®, West Lafayette, IN, USA)—as reference electrode, and a Pt wire as counter electrode. To obtain a constant surface area for Ni electrodeposition, the PGEs were isolated with Teflon band, leaving a controlled length of 20 mm in contact with the electrolyte solution. Before use, the PGEs were cleaned in ethanol using an ultrasound bath and finally rinsed with distilled water.

Table 1. The composition of the citrate and Watts baths used for the deposition of nickel.

Bath Type	Composition	Concentration	pH
Citrate	$Na_3C_6H_5O_7 \cdot 5H_2O$ $NiSO_4 \cdot 6\,H_2O$	0.3 mol L^{-1} 0.2 mol L^{-1}	8
Watts	$NiSO_4 \cdot 6\,H_2O$ $NiCl_2 \cdot 6\,H_2O$ H_3BO_3	300 g L^{-1} 45 g L^{-1} 30 g L^{-1}	2

Constant cathodic current pulses of 25 and 50 mA cm^{-2} were applied using an Autolab 302N potentiostat/galvanostat (Metrohm Autolab, Utrecht, The Netherlands) with the following deposition parameters: on time t_{on} = 0.1 s, off time t_{off} = 0.5 s, and 500 cycles, resulting in a duty cycle of 16.67%. Several electrodes were prepared by electrodeposition at three different temperatures, namely 20, 40, and 60 °C, while keeping the pulse parameters as stated. The fabrication process is described in Figure 1.

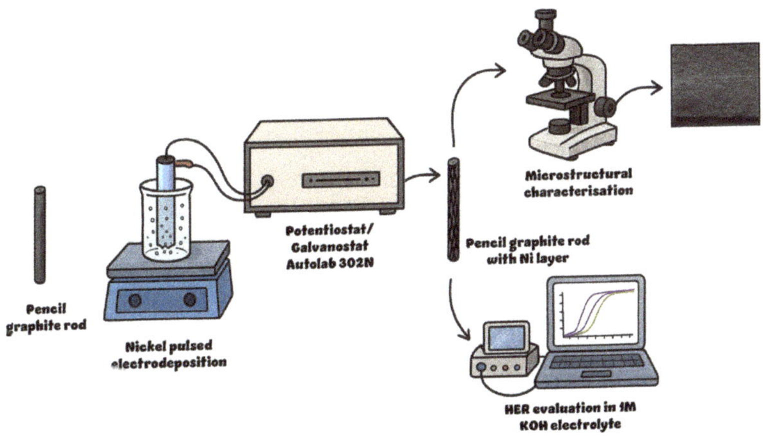

Figure 1. Schematic representation of the electrode fabrication and testing.

The citrate bath was prepared using nickel sulfate hexahydrate ($NiSO_4 \cdot 6\ H_2O$, Chimreactiv, p.a.) and trisodium citrate pentahydrate ($Na_3C_6H_5O_7 \cdot 5H_2O$, Chimreactiv, p.a.) with the composition given in Table 1. After preparation, the pH was adjusted to 8 by dropwise addition of ammonia.

In the citrate bath, trisodium citrate acts as a complexing agent, forming nickel–citrate complexes [36] that influence the nucleation and growth of nickel deposits. This results in smoother, more uniform Ni layers with finer crystal structure than those obtained from the Watts bath [37]. Nickel electrodeposited from citrate baths typically exhibits a compact, globular surface and nanocrystalline microstructure [28,38]. In the Watts bath, nickel sulfate hexahydrate ($NiSO_4 \cdot 6\ H_2O$, Chimreactiv, p.a.) and nickel chloride hexahydrate ($NiCl_2 \cdot 6\ H_2O$, Merck, EMSURE®, ACS) provide the necessary nickel ions for the cathodic deposition, while boric acid (H_3BO_3 Merck, EMSURE®, ACS, ISO, Reag. Ph Eur) acts as a buffering agent to ensure the acidic environment during the electrodeposition process.

2.2. Physico-Chemical Characterization of Ni-PGE

The structure and morphology of Ni-PGE were studied using field emission scanning electron microscopy (FE-SEM) on a QUANTA FEG 250 microscope (Quanta FEG 250, FEI, Hillsboro, OR, USA), with the Everhart–Thornley detector (ETD) for secondary electrons (SE). Images were taken in a high vacuum mode with an accelerating voltage of 10 kV and a working distance of 10 mm. The elemental composition was determined by energy dispersive X-ray analysis (EDAX with Apollo SSD detector, EDAX Inc., Mahwah, NJ, USA).

2.3. Performance of Ni-PGE for Hydrogen Evolution Reaction

The performance of Ni-PGE for HER was evaluated using a combination of electrochemical methods, including linear sweep voltammetry (LSV), chronopotentiometry (CP), and electrochemical impedance spectroscopy (EIS). All electrochemical tests were carried out on an Autolab 302 N potentiostat/galvanostat, in a three-electrode configuration cell, with Ni-PGE as working electrode, a silver—silver chloride (Ag/AgCl in 3 M KCl)—as reference electrode and a Pt wire as counter electrode. The measured electrode potentials versus the reference were converted to the reversible hydrogen electrode (RHE) using the formula $E_{RHE} = E_{Ag/AgCl} + 0.059\ pH + E^{o}_{Ag/AgCl}$, where $E_{Ag/AgCl}$ is the measured potential, and $E^{o}_{Ag/AgCl}$ is the potential of the Ag/AgCl electrode (0.210 V for a 3 M KCl solution). The test solution was 1 M KOH (Merck, EMSURE®, p.a.).

LSV measurements were recorded in 1 M KOH at a scan rate of 10 mV s^{-1} to obtain the Tafel plots and calculate the kinetic parameters: b is the Tafel slope, and i_0 is the exchange current density. CP measurements were conducted in 1 M KOH at a current density of 10 and 100 mA cm^{-2} for 2 h to investigate HER overpotentials and Ni-PGE stability. EIS measurements were carried out over a frequency range of 100 kHz to 10 mHz with an AC voltage amplitude of 10 mV rms. For each spectrum, 60 points were collected using a logarithmic distribution of 10 points per frequency decade. Impedance spectra for each Ni-PGE were measured at five different HER overpotentials ranging from −50 mV to −250 mV. The experimental EIS data were modeled to an equivalent electrical circuit (EEC) using a complex non-linear least-squares Levenberg–Marquardt procedure with the ZView 3.0 software. All electrochemical measurements were repeated on a nickel wire (ϕ = 1 mm, 99.2% Ni), and the HER performance of Ni-PGEs was compared to that of a reference Ni wire electrode.

3. Results and Discussion

3.1. *Pulsed Current Electrodeposition of Ni-PGE*

To investigate the effect of bath type and electrodeposition parameters—pulsed current density and temperature—on the electrocatalytic activity for HER, several electrodes were prepared by varying the parameters according to the data presented in Table 2.

Table 2. Electrodeposition parameters used for preparation of Ni-PGE.

Electrode	Pulsed Current Density (mA cm^{-2})	Temperature (°C)	Electrodeposition Conditions
Ni-PGE-W1	50	25	Watts bath pH = 2
Ni-PGE-W2	50	40	
Ni-PGE-C1	25	25	Citrate bath pH = 8
Ni-PGE-C2	25	40	
Ni-PGE-C3	25	60	
Ni-PGE-C4	50	25	
Ni-PGE-C5	50	40	
Ni-PGE-C6	50	60	

In all cases, the pulse parameters were maintained similarly, as described in the Materials and Methods Section. For the Watts bath, a pulsed current density of 25 mA cm^{-2} was not adequate to achieve deposition; therefore, only deposits obtained at 50 mA cm^{-2} were further investigated.

3.2. *Characterization of Ni-PGE*

Figure 2 presents FE-SEM images of Ni deposits obtained from an acidic Watts bath at a current density of 50 mA cm^{-2} and two different temperatures: 25 and 40 °C. The EDAX elemental analysis results are shown as insets in Figure 2b,d. The observed nickel deposit morphology is consistent with nickel electrodeposition from the Watts bath occurring simultaneously with the cathodic HER.

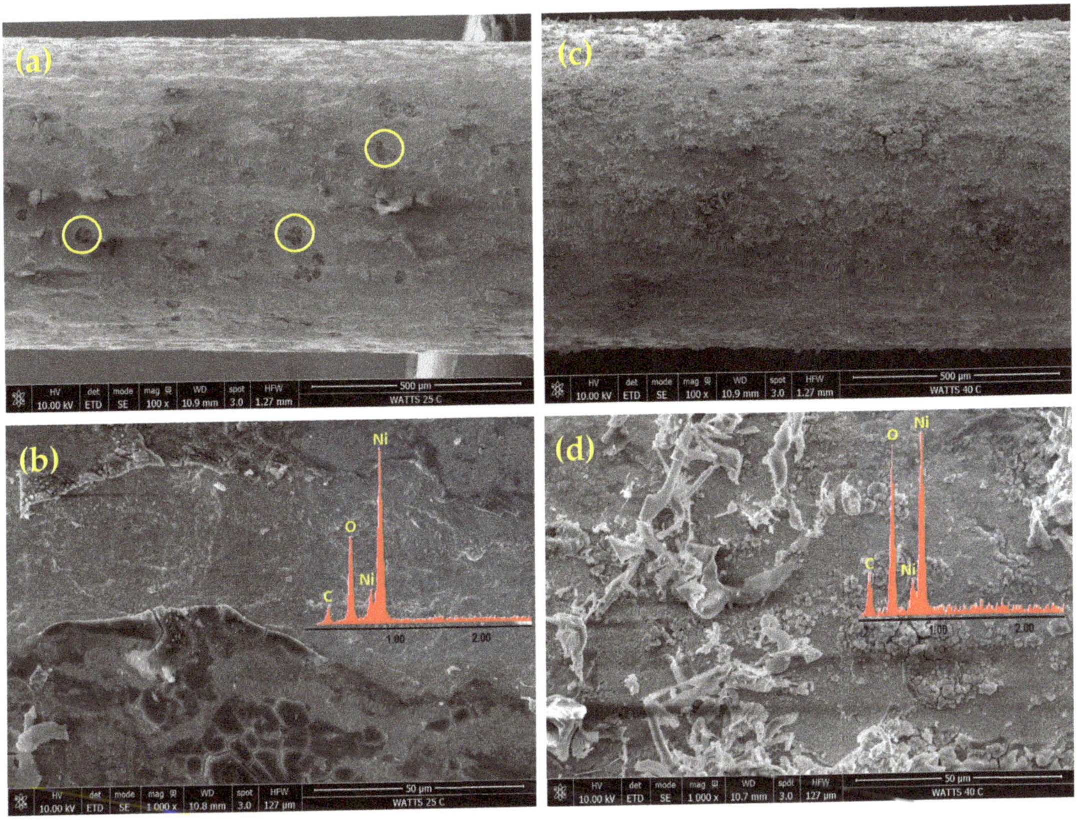

Figure 2. FE-SEM micrographs of nickel deposits on PGE obtained by pulsed current electrodeposition from the Watts bath at 50 mA cm^{-2} current density and the two different temperatures of 25 °C and 40 °C: Ni-PGE-W1 (**a**,**b**) and Ni-PGE-W2 (**c**,**d**). Insets show the results of EDAX elemental analysis.

The lower-magnitude FE-SEM images from Figure 2a,c reveal a heterogeneous, relatively compact structure, indicating a low specific surface area. The deposit obtained at 25 °C (Figure 2a) shows a rough texture with visible irregularities and a large number of circular defects (marked as yellow circles) resulting from hydrogen bubbles entrapment or shielded regions where nickel deposition was hindered by gas evolution. When the temperature is increased to 40 °C (Figure 2c), the deposition rate increased, resulting in more uniform coverage, but at the same time, dendritic structures and filament-like growths began to form. In the Watts bath, increasing temperature enhanced the diffusion of Ni^{2+} ions towards the cathode and favored rapid growth at existing nucleation sites rather than the formation of new sites. This explains the formation of dendritic or filament-like structures, particularly under high current density conditions. At increased magnification, the FE-SEM images in Figure 2b,d show a layered structure, with a compact base layer and globular morphology. At 25 °C (Figure 2b), numerous cracks are visible in the top layer, caused by internal stresses during deposition. At 40 °C (Figure 2d), a finer grain structure is observed covered by filaments with a dendritic structure.

Figure 3 shows selected FE-SEM images of the Ni deposits obtained from the alkaline citrate bath at 40 °C and two different current densities: 25 and 50 mA cm^{-2}.

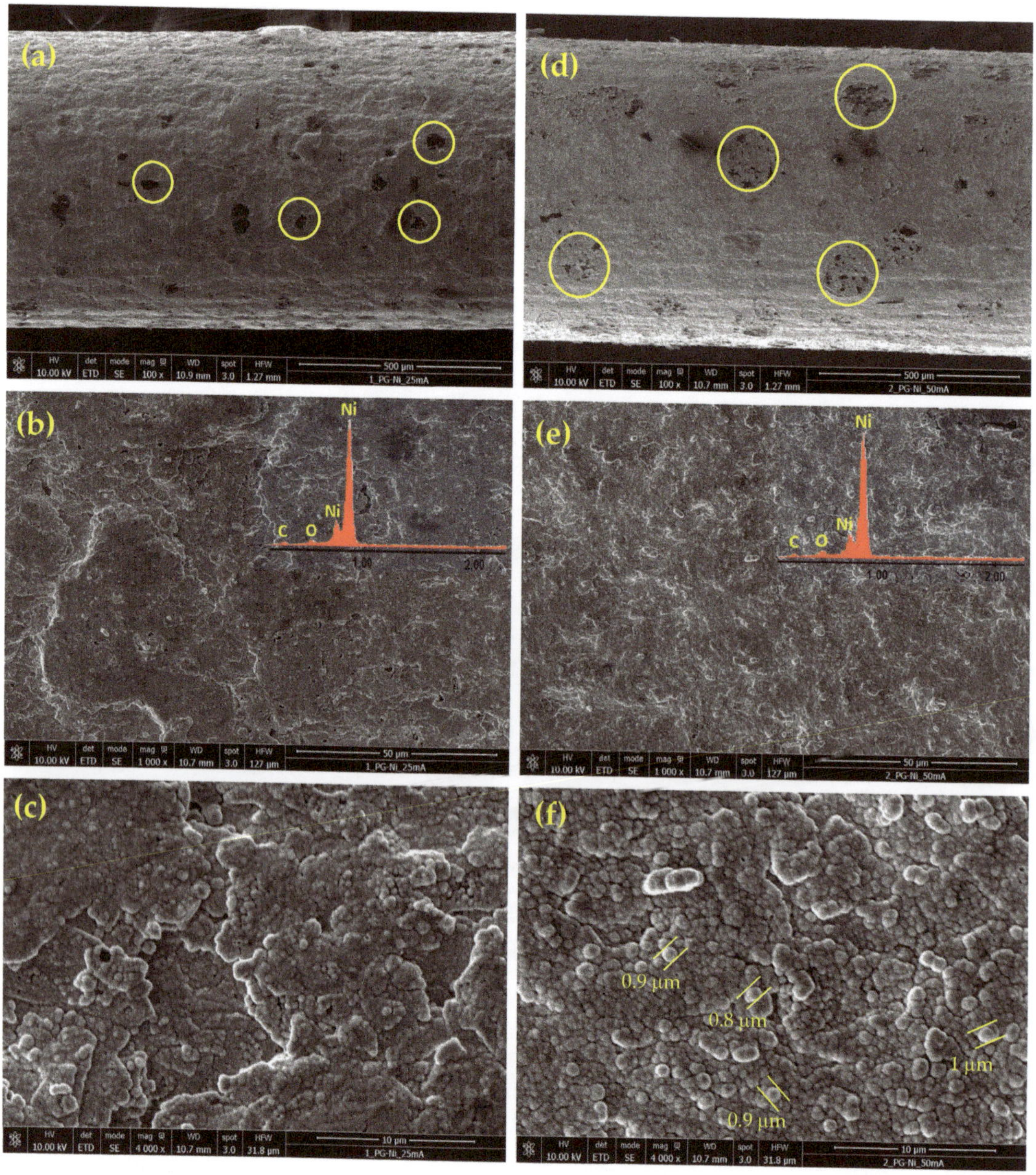

Figure 3. FE-SEM micrographs of nickel deposits on PGE obtained by pulsed current electrodeposition from the citrate bath at 40 °C and the two different current densities of 25 and 50 mA cm^{-2}: Ni-PGE-C2 (**a–c**) and Ni-PGE-C5 (**d–f**). Insets show the results of EDAX elemental analysis.

Similar to the Watts bath, nickel electrodeposition from citrate bath occurs simultaneously with HER. However, in alkaline solution, the deposited nickel acts as an electrocatalyst for HER. Consequently, the FE-SEM images in Figure 3a,d reveal more frequent and larger shielded areas where nickel deposition was hindered due to hydrogen bubbles formation (yellow circles), as compared to the Watts bath. As the deposition current density increased, the current efficiency for nickel deposition decreased due to the intensification of the secondary reaction, and the spots corresponding to hydrogen bubbles became even larger. Simultaneously, higher current density produces a more uniform deposition structure. Comparing Figure 3b,e, it is evident that at low current density, more pores and voids formed, whereas at high current density, the deposit became more uniform and homogeneous. At higher magnification, the nickel deposits in Figure 3f exhibit a homogeneous, globular morphology, with structures approximately 1 μm in size and characteristic of citrate-based alkaline baths [38]. This morphology provides a higher specific surface area compared to deposits from the acidic Watts bath. Unlike the Watts bath, the presence of complexing agents in the citrate bath regulated the availability of Ni^{2+} ions at the cathode surface. This ensured a more controlled and uniform nucleation process, even at elevated temperatures, reducing the tendency to anisotropic growth. As a result, nickel deposits from citrate baths typically display finer grains and a nanocrystalline structure, leading to an increased specific surface area.

The EDAX analysis confirmed the presence of nickel, carbon, and oxygen for all analyzed samples. The carbon originated from the graphite substrate, while the oxygen content was higher in nickel deposits from the acidic Watts bath compared to those from the alkaline citrate bath. This suggests a pronounced incorporation of oxidized species, such as nickel hydroxide or nickel oxide, in deposits obtained from the Watts bath. In contrast, the alkaline citrate bath contained citrate ligands, acting as complexing agents, stabilizing nickel ions, and reducing their direct hydrolysis or oxide formation, leading to lower oxygen incorporation in the deposit.

3.3. *Electrocatalytic Activity of Ni-PGE Towards Hydrogen Evolution Reaction*

Figure 4 shows the performance of Ni-PGE for HER in alkaline solution, which was evaluated based on linear potentiodynamic polarization curves, Tafel plots, the overpotential value required to reach a current density of 10 and 50 mA cm^{-2}, and stability tests at a current density of 10 and 100 mA cm^{-2}, respectively. The polarization curves obtained for Ni deposits from the Watts (Figure 4a) and citrate baths (Figure 4c) reveal that unlike pencil graphite, which showed no electrocatalytic activity for HER, all the Ni-PGE showed good electrocatalytic activity. For both types of Ni deposits obtained from the Watts and citrate baths, increasing the deposition temperature from 25 to 40 °C had a favorable effect on the performance of the electrodes at both 25 and 50 mA cm^{-2} applied pulsed current densities. Further increasing the deposition temperature to 60 °C for the citrate bath negatively affected the electrocatalytic activity. In conclusion, we established that the optimal deposition temperature is 40 °C.

Tafel plots for HER of Ni-PGEs from the Watts (Figure 4b) and citrate baths (Figure 4d) allowed for a quantitative assessment of the electrocatalytic activity for HER, using the kinetic parameters (i_o—exchange current density; b—Tafel slope). The values of the kinetic parameters, together with the overpotential necessary to reach current densities of 10, 50, and 100 mA cm^{-2}, respectively, are summarized in Table 3.

Table 3. Values of kinetic parameters for HER on Ni and Ni-PGE in 1 M KOH solution.

Electrode	i_o (mA cm^{-2})	b (mV dec^{-1})	η @ 10 mA cm^{-2}	η @ 50 mA cm^{-2}	η @ 100 mA cm^{-2}
Ni	0.28	−133	−200	−301	−380
Ni-PGE-W1	0.63	−176	−269	−384	-
Ni-PGE-W2	0.67	−122	−213	−327	−396
Ni-PGE-C1	0.60	−218	−230	−336	−420
Ni-PGE-C2	0.61	−175	−226	−326	−396
Ni-PGE-C3	0.48	−204	−251	−370	-
Ni-PGE-C4	0.60	−218	−219	−335	−422
Ni-PGE-C5	0.93	−208	−210	−314	−386
Ni-PGE-C6	0.53	−243	−247	−356	-

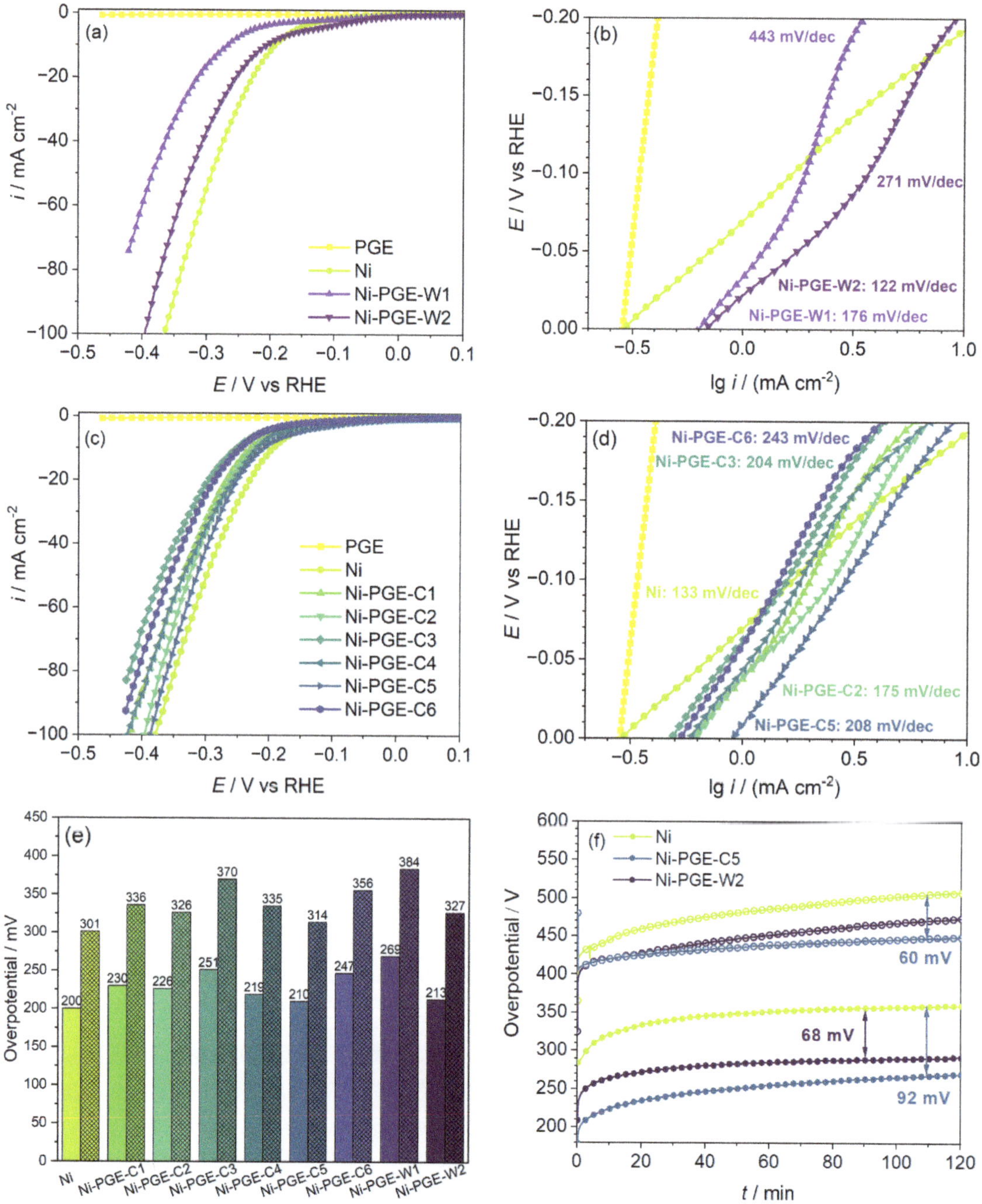

Figure 4. Performance of Ni-PGE for HER in 1 M KOH: polarization curves at 10 mV s^{-1} scan rate and Tafel plots for Ni-PGE obtained by pulsed current electrodeposition from (**a**,**b**) Watts bath and (**c**,**d**) citrate bath; (**e**) overpotential required to attain a current density of 10 and 50 mA cm^{-2} for Ni and Ni-PGE; (**f**) chronopotentiometric stability test of Ni, Ni-PGE-W2, and Ni-PGE-C5 at a current density of 10 mA cm^{-2} (closed symbols) and 100 mA cm^{-2} (open symbols). Polarization curves were *iR*-corrected using the ohmic resistances obtained from impedance data.

The exchange current density i_0 is a parameter used to evaluate the reaction rate at the equilibrium potential. According to the data presented in Table 3, all Ni-PGEs have higher exchange current density values than Ni. However, these values are apparent, as they depend on the specific surface area of the electrodes. Another essential parameter for assessing the electrochemical performance of electrocatalysts away from equilibrium is the overpotential required to achieve a specific current density. Typically, the overpotentials necessary to reach current densities of 10 or 100 mA cm^{-2} are reported. Figure 4e provides an overview of the overpotentials needed to attain 10 and 50 mA cm^{-2} for all Ni-PGEs. The Ni wire electrode, used as reference, exhibited an overpotential of approximately 200 mV at 10 mA cm^{-2} and 300 mV at 50 mA cm^{-2}. In comparison, Ni-PGE electrodes generally showed slightly higher overpotentials, depending on the current density, bath composition, and temperature applied during electrodeposition of nickel. Within the citrate bath series, a higher electrodeposition current density resulted in modified electrodes with lower HER overpotential, indicating improved catalytic activity. Additionally, increasing the deposition temperature from 25 °C to 40 °C reduced the HER overpotential, enhancing performance. However, a further increase in temperature during electrodeposition to 60 °C resulted in modified electrodes with higher HER overpotentials compared to those obtained at 25 and 40 °C. This suggests the existence of an optimal temperature range (e.g., around 40 °C) where nickel electrodeposition yields the most favorable microstructure and electrochemical performance. Beyond this range, further increase in temperature negatively affected HER efficiency. Among these electrodes, Ni-PGE-C5 exhibited the lowest overpotential (210 mV at 10 mA cm^{-2} and 314 mV at 50 mA cm^{-2}), making it the most effective catalyst in this series. Conversely, Ni-PGE-C3 displayed the highest overpotential (251 mV at 10 mA cm^{-2} and 370 mV at 50 mA cm^{-2}), indicating reduced HER efficiency. For electrodes obtained from the Watts bath, Ni-PGE-W1 demonstrated the highest overpotential overall, indicating that room temperature deposition from the Watts bath negatively affects HER performance. However, Ni-PGE-W2 performed relatively better, with an overpotential of 213 mV at 10 mA cm^{-2} and 327 mV at 50 mA cm^{-2}, making it comparable to Ni-PGE-C5. Figure 4f presents the stability tests conducted at a constant current density of 10 and 100 mA cm^{-2} for the best-performing electrodes. The results show a marked increase in overpotential at 10 mA cm^{-2} during the first 20 min, followed by a slower, more gradual increase before stabilizing after approximately 2 h. Among the tested materials, Ni exhibited the highest overpotential for HER at around 360 mV. In contrast, Ni-PGE-W2 achieved a reduction of approximately 68 mV, while Ni-PGE-C5 exhibited an even greater decrease of about 92 mV, indicating superior catalytic performance. The trend in overpotentials was maintained even at a higher current density of 100 mA cm^{-2}, although the difference between Ni and Ni-PGE-C5 is smaller. The results of the stability tests indicate a different order of electrocatalytic activity than that derived from the polarization curves. It is interesting to note that, in the polarization curves, the initial overpotential for Ni is lower (200 mV) compared to Ni-PGE-C5 (210 mV) and Ni-PGE-W2 (213 mV). This suggests that under short-term conditions, Ni appears slightly better at initiating HER. On the contrary, the stability test showed that over time, Ni exhibited the highest overpotential (360 mV), while Ni-PGE-C5 and Ni-PGE-W2 maintained significantly lower values (270 mV and 290 mV, respectively). The discrepancies can be explained by the differences between the two methods. The polarization curves reflect instantaneous performance, and they do not capture long-term stability or electrode degradation. The stability test highlights how the electrodes perform under prolonged electrochemical operation. The increased overpotential of Ni suggests surface passivation or catalyst site deactivation during long-term tests. Ni-PGE-C5 and Ni-PGE-W2 exhibited lower overpotentials, implying better electrocatalytic activity for HER. Although Ni initially showed a lower overpotential in the polarization curves, it experienced significantly higher overpotentials over time, making it less effective for prolonged HER applications. In contrast, Ni-PGE-C5, despite having a slightly higher initial overpotential, outperformed Ni in stability, making it the superior electrode for long-term HER performance.

According to the data presented in Table 3, the Tafel slope obtained for Ni is consistent with values reported in previous studies [39,40] and closely approaches the theoretical value of -118 mV dec^{-1}. This suggests that the charge transfer step (Volmer reaction) is likely the rate-determining step on Ni. In contrast, the Ni-PGE electrodes exhibited higher Tafel slopes, ranging from -175 to -243 mV dec^{-1}. Tafel slopes around -200 mV dec^{-1} are commonly reported for Ni-based electrocatalysts under similar conditions [41,42]. Notably, a change in Tafel slope was observed at overpotentials exceeding -100 mV, particularly for Ni-PGE samples deposited from the Watts bath, indicating possible changes in the reaction mechanism. Additionally, both the exchange current density (i_o) and Tafel slope (b) were influenced by the electrodeposition temperature, with the highest i_o and lowest b values consistently observed at 40 °C across all series.

The performance of Ni-PGEs was further investigated using electrochemical impedance spectroscopy. To gain deeper insight into their behavior and underlying HER mechanism, EIS measurements were carried out at different overpotential values. The results are presented in Figure 5 as complex plane Nyquist plots (Figure 5a) and impedance magnitude versus frequency plots (Figure 5b), both recorded at the highest overpotential of -250 mV. Additionally, the dependence of charge transfer resistance (Figure 5c) and double-layer capacitance (Figure 5d) versus overpotential is shown. The Nyquist plots for nickel exhibit a single semicircle across all measured overpotentials (Figure S1). In contrast, the Nyquist plots for Ni-PGE display an overpotential dependent shape, exemplified for Ni-PGE-C5 in Figure S3. At low overpotentials (bellow -100 mV), the complex plane plots feature a depressed semicircle at high frequencies and either an overlapping semicircle or a tail at low frequencies. At higher overpotentials (above -100 mV), only a single semicircle is observed. The presence of a single time constant in the phase angle Bode plots for Ni can be observed in Figure S2, while the appearance of two time constants at low overpotentials for Ni-PGE-C5 is evident in Figure S4. This behavior is consistent with the differences observed in the Tafel slopes; i.e., Ni exhibits a single Tafel slope, while for Ni-PGEs, the Tafel slope depends on the overpotential. To model this behavior, two different equivalent electric circuits were employed, as given in Figure 5e,f. Both EECs comprise the solution resistance (R_S) in series with a parallel connection of the double-layer capacitance (C_{dl}, represented as a constant phase element CPE) and the charge transfer resistance (R_{ct}). For low overpotentials, a second parallel branch composed of R_{ads} and C_{ads} was added to model the low-frequency behavior associated with adsorption phenomena. Here, R_{ads} represents the mass transfer resistance of the adsorbed hydrogen species, while C_{ads} accounts for the pseudo-capacitance of the adsorbed hydrogen layer [43]. The high-frequency semicircle in the Nyquist plot corresponds to double-layer charging and fast charge transfer (C_{dl}-R_{ct}), whereas the low-frequency semicircle or tail is attributed to mass transport of adsorbed hydrogen (C_{ads}-R_{ads}). The obtained impedance spectra for Ni-PGE at low overpotentials align well with this mechanism, revealing a depressed semicircle at high frequencies followed by an overlapping semicircle or a tail at low frequencies. The diameter of both semicircles is overpotential-dependent, confirming they are both related to HER kinetics.

Several studies have demonstrated that at lower cathodic overpotentials, two relaxation frequencies in impedance spectra allow separation between different steps of the HER mechanism, highlighting the contribution of hydrogen adsorption processes in this potential range [41,42,44,45]. In the Nyquist plots, the low-frequency semicircle is ascribed to the charge transfer process, e.g., Volmer reaction in the HER mechanism, while the high-frequency semicircle could be associated with the mass transfer processes of the adsorbed species at the cathode [43,46].

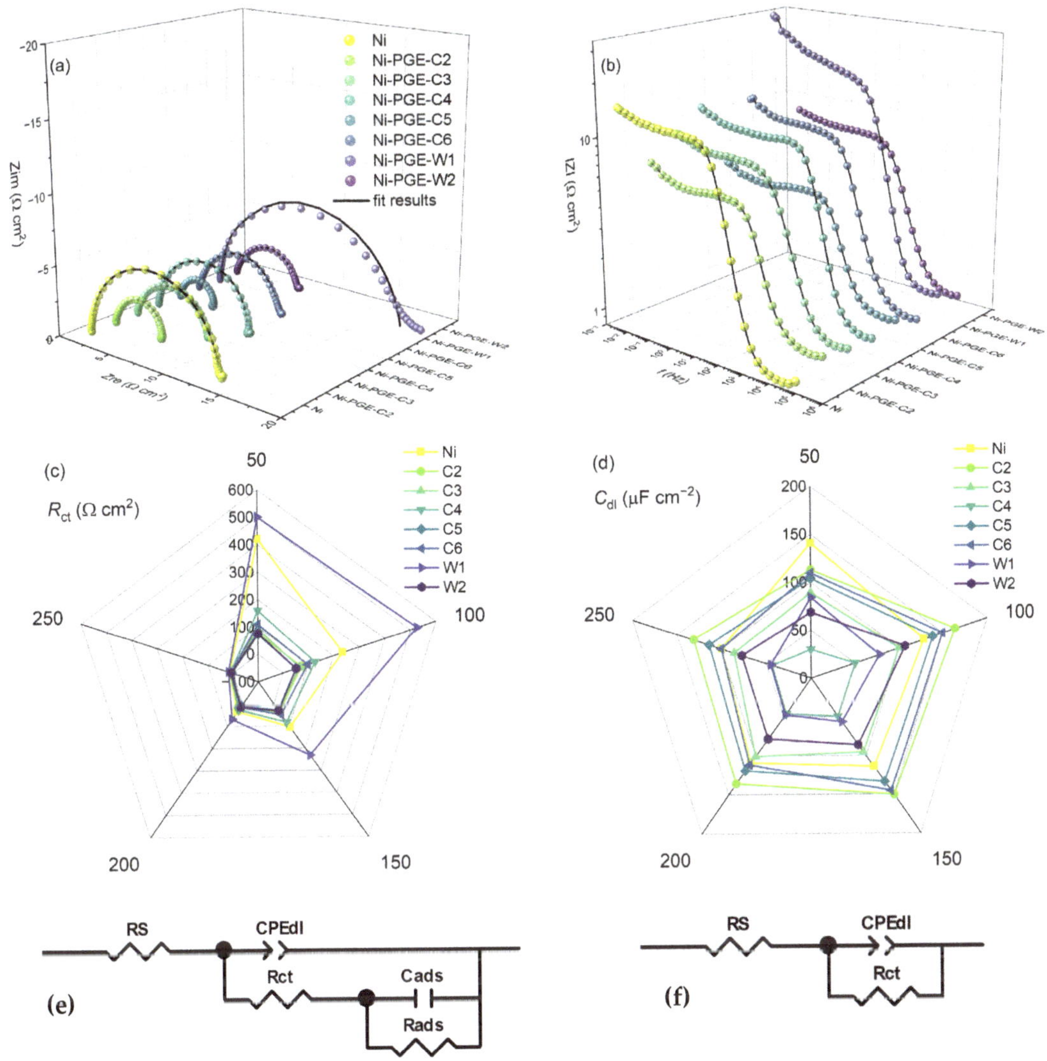

Figure 5. Electrochemical impedance data of Ni and Ni-PGE in 1 M KOH measured at an overpotential of −250 mV. (**a**) Nyquist and (**b**) Bode plots; (**c**) charge transfer resistance and (**d**) double-layer capacitance as a function of overpotential; (**e**) equivalent electrical circuit used to model the impedance data at low and (**f**) high overpotentials.

The Nyquist plots in Figure 5a reveal the presence of a single depressed semicircle, whose diameter corresponds to the charge transfer resistance. It can be observed that all Ni-PGE samples deposited at 40 °C, i.e., Ni-PGE-C2, Ni-PGE-C5, and Ni-PGE-W2, consistently show lower-diameter semicircles compared to samples deposited at different temperatures and Ni. These samples also present the lowest absolute impedance values in the Bode plots illustrated in Figure 5b.

The impedance parameters obtained by fitting the experimental data obtained at the highest overpotential of −250 mV, using the EEC given in Figure 5f, are summarized in Table 4. The impedance parameters obtained at lower overpotentials are given in the Supplementary Material in Table S1 (η = −200 mV), Table S2 (η = −150 mV), Table S3 (η = −100 mV), and Table S4 (η = −50 mV), respectively, along with their associated errors and the goodness-of-fit indicators, represented by chi-square values. The low errors and small chi-square values indicate a high-quality fit between the experimental data and the proposed equivalent circuit models.

Table 4. Impedance parameters for HER on Ni and Ni-PGE in 1 M KOH solution at the highest overpotential of −250 mV, obtained by fitting the impedance data to the model in Figure 5f.

Electrode	R_S (Ω cm^2)	CPE-T (F cm^{-2} s^{n-1})	n	R_{ct} (Ω cm^2)	C_{dl} (μF cm^{-2})	Chi^2
Ni	1.3 (0.7%)	1.49×10^{-4} (6.2%)	0.95 (0.8%)	5.0 (0.8%)	99.8	3.36×10^{-3}
Ni-PGE-C2	1.1 (0.7%)	2.99×10^{-4} (6.1%)	0.90 (0.8%)	4.7 (0.9%)	131.4	3.39×10^{-3}
Ni-PGE-C3	1.0 (0.7%)	3.77×10^{-4} (4.6%)	0.84 (0.7%)	6.9 (0.8%)	85.4	1.97×10^{-3}
Ni-PGE-C4	1.0 (0.8%)	7.36×10^{-5} (4.9%)	0.95 (0.6%)	8.8 (0.8%)	43.5	2.00×10^{-3}
Ni-PGE-C5	1.1 (0.6%)	2.47×10^{-4} (6.2%)	0.91 (0.8%)	3.5 (0.9%)	113.9	1.96×10^{-3}
Ni-PGE-C6	0.9 (0.9%)	3.82×10^{-4} (5.2%)	0.85 (0.7%)	8.0 (1.0%)	100.1	2.95×10^{-3}
Ni-PGE-W1	1.1 (0.8%)	8.63×10^{-5} (4.0%)	0.93 (0.5%)	16.7 (0.7%)	44.8	2.50×10^{-3}
Ni-PGE-W2	0.9 (0.8%)	3.74×10^{-4} (4.9%)	0.85 (0.7%)	6.3 (0.8%)	76.7	2.12×10^{-3}

The double-layer capacitance was calculated from the constant phase element parameters, whose impedance is given by

$$Z_{CPE} = T^{-1}(j\omega)^{-n} \tag{1}$$

by applying Brug's relationship [47]:

$$C_{dl} = T^{1/n}\,(R_S{}^{-1} + R_{ct}{}^{-1})^{(n-1)/n} \tag{2}$$

where T is a frequency independent parameter (F cm^{-2} s^{n-1}) related to the capacitance, j is the imaginary unit, ω is the angular frequency (rad s^{-1}), and n is a dimensionless parameter between 0 and 1 that describes the constant phase angle of CPE and is equal to $-(n*90)$ [48].

Based on the impedance data presented in Table 4, it can be observed that Ni-PGE-C5 shows the lowest R_{ct} value, indicating the most effective charge transfer kinetics among the tested electrodes. Notably, increasing the deposition temperature for both the citrate and Watts baths led to an increase in the surface roughness, as evidenced by the decrease in parameter n, which is typically associated with surface inhomogeneities. However, the trends in double-layer capacitance (C_{dl}) and charge transfer resistance (R_{ct}) did not follow this behavior. Instead, the data suggest that a deposition temperature of 40 °C yields optimal electrocatalytic performance, characterized by the highest C_{dl} and the lowest R_{ct}.

The results presented in Figure 5c and Tables S1–S4 show that R_{ct} values decrease with increasing overpotential. Across all overpotentials, the samples obtained by electrodeposition at 50 mA cm^{-2} and 40 °C (Ni-PGE-C5 and Ni-PGE-W2) consistently exhibited the lowest R_{ct} values, indicating faster charge transfer kinetics. Furthermore, comparing the samples obtained by electrodeposition at 50 mA cm^{-2} and 25 °C from the citrate and Watts baths (Ni-PGE-C4 and Ni-PGE-W1) reveals that the electrocatalyst obtained from the citrate bath always presented lower R_{ct} values than its Watts bath counterpart, highlighting the favorable effect of the citrate bath on electrocatalyst performance. Analysis of the double-layer capacitance values presented in Figure 5d and Tables S1–S4 indicates that C_{dl} decreased with increasing overpotential. This can be explained by the reduction in the electrochemical active surface area caused by gas bubble accumulation and blocking effects resulting from intensified hydrogen evolution at more negative overpotentials [49,50]. The highest C_{dl} values were observed for Ni-PGE-C2 (electrodeposited at 25 mA cm^{-2} and 40 °C); however, these values do not correspond to the lowest R_{ct} values. In contrast, Ni-PGE-C4 (electrodeposited at 50 mA cm^{-2} and 25 °C) showed the lowest C_{dl} values and were equivalent to the lowest active surface area.

For a clearer overview on Ni-PGE electrocatalysts performance for HER, Table 5 gives a comparison of kinetic parameters related to HER for the Ni-PGE electrodes developed in this work, alongside different literature-reported Ni-based electrocatalysts obtained by potentiostatic or galvanostatic electrodeposition from different types of baths.

Table 5. Comparative data on the kinetic parameters for HER and overpotential at 10 and 100 mA cm^{-2} for different Ni-based electrocatalysts.

Electrode	Deposition Bath/pH	Testing Solution	i_o (mA cm^{-2})	b (mV)	η @ 10 mA cm $^{-2}$	η @ 100 mA cm $^{-2}$	Reference
Ni-PGE-C5	Citrate/pH = 8	1 M KOH	0.93	−208	−210	−392	This work
Ni-PGE-W2	Watts/pH = 2	1 M KOH	0.67	−122	−213	−432	This work
Ni thin film	Citrate/pH = 1.47	1 M NaOH	1.98	−118	−231	over −400 *	[28]
NiMo-1	Citrate/pH = 10.5	1 M NaOH	1.0	−192	−185 *	−384	[50]
NiMo30%	Citrate/pH = 9	1 M KOH	0.487	−124	−180	n/a	[51]
NiMo	Citrate/pH = 10	1 M KOH	0.93	−152	−154	n/a	[52]
$Ni/NiO/Ni(OH)_2$	$NiCl_2$/acid	0.1 M KOH	0.16	−118	−210	n/a	[53]
$Ni/NiO/Ni(OH)_2$	$NiCl_2$/acid	1 M KOH	0.057	−88	−197	n/a	[53]
Ni/NiO_x	Ni oxidation	1 M NaOH	0.0365	n/a	n/a	n/a	[54]
Ni	Watts/pH = 3.5	1 M NaOH	0.012	−93	−263	−350	[55]

* estimated from polarization curves.

The Ni-PGE-C5 electrode, obtained via pulsed current electrodeposition from a citrate bath at 50 mA cm^{-2} and 40 °C, demonstrated excellent electrocatalytic activity towards HER. It achieved a high exchange current density comparable to that of more complex electrocatalysts, such as NiMo alloy obtained by galvanostatic electrodeposition at 30 °C from a citrate bath with pH around 10 [50,52] or by pulsed current electrodeposition at 40 °C from a citrate bath with pH = 9 [51]. Despite having slightly lower i_o values and a higher Tafel slope than Ni thin film deposited at a constant potential from a citrate bath at pH = 1.47 under the influence of an external magnetic field [28], it compensated by exhibiting lower overpotentials at current densities of 10 and 100 mA cm^{-2}. Furthermore, in terms of exchange current density, Ni-PGE-C5 significantly outperformed Ni deposited from the Watts bath at a constant current density of 50 mA cm^{-2} and room temperature [55] as well as nickel composite nanoparticles consisting of Ni, NiO, and $Ni(OH)_2$ prepared by cyclic voltammetry [53]. These results highlight the potential of Ni-PGE-C5 as a cost-effective and efficient HER electrocatalyst for alkaline water electrolysis, particularly considering its simple preparation method.

4. Conclusions

Nickel electrodeposition from both the acidic Watts and alkaline citrate baths was performed by pulsed current method on pencil graphite substrate at different current densities and temperatures to investigate the effect on the electrocatalytic performance for HER in 1 M KOH solution. The obtained results showed the following:

(1) The morphology of the Ni deposit strongly depends on the type of bath used for electrodeposition, revealing a heterogeneous, layered structure with dendritic and

filament-like growths for the Watts bath and a homogeneous globular morphology, with structures approximately 1 μm in size for the citrate bath;

(2) Linear polarization curves for HER in 1 M KOH solution indicate that increasing the electrodeposition temperature from 25 to 40 °C improves the performance of Ni-PGE obtained from both the Watts and citrate baths at both applied pulsed current densities. Further increasing the electrodeposition temperature to 60 °C has a negative effect, resulting in Ni electrocatalysts with lower HER performance;

(3) Kinetic parameters extracted from the Tafel plots demonstrate higher exchange current density values for all Ni-PGE electrocatalysts as compared to metallic Ni, which is attributed to an increased active surface area;

(4) Chronopotentiometric measurements of HER overpotentials indicate that Ni-PGE-C5 electrocatalyst obtained from the citrate bath at 40 °C and 50 mA cm^{-2} achieves the best HER performance over time, showing an important overpotential reduction compared to metallic Ni;

(5) EIS analysis revealed the lowest charge transfer resistance for Ni-PGE-C5, indicating enhanced HER kinetics and better electrocatalytic properties.

These results highlight the optimal electrodeposition parameters regarding the bath type, temperature, and current density required to design efficient HER electrocatalysts for applications in alkaline water electrolysis.

Supplementary Materials: The following supporting information can be downloaded at https://www.mdpi.com/article/10.3390/ma18122775/s1. Figure S1: Nyquist plots of Ni in 1 M KOH measured at different overpotentials; Figure S2: Bode plots of Ni in 1 M KOH measured at different overpotentials; Figure S3: Nyquist plots of Ni-PGE-C5 in 1 M KOH measured at different overpotentials; Figure S4: Bode plots of Ni-PGE-C5 in 1 M KOH measured at different overpotentials; Table S1: Impedance parameters for HER on Ni and Ni-PGE in 1 M KOH solution at an overpotential of −200 mV obtained by fitting the impedance data to the model in Figure 5f; Table S2: Impedance parameters for HER on Ni and Ni-PGE in 1 M KOH solution at an overpotential of −150 mV obtained by fitting the impedance data to the model in Figure 5f; Table S3: Impedance parameters for HER on Ni and Ni-PGE in 1 M KOH solution at an overpotential of −100 mV obtained by fitting the impedance data to the model in Figure 5e; Table S4: Impedance parameters for HER on Ni and Ni-PGE in 1 M KOH solution at an overpotential of −50 mV obtained by fitting the impedance data to the model in Figure 5e.

Author Contributions: Conceptualization, A.K. and R.M.; methodology, A.K. and R.M.; formal analysis, A.K. and R.B.; investigation, R.C., R.M. and R.B.; resources, A.K.; data curation, A.K. and R.B.; writing—original draft preparation, A.K., R.M. and R.B.; writing—review and editing, A.K., R.M., R.B. and R.C.; visualization, A.K. and R.M.; supervision, A.K. All authors have read and agreed to the published version of the manuscript.

Funding: This research received no external funding.

Institutional Review Board Statement: Not applicable.

Informed Consent Statement: Not applicable.

Data Availability Statement: The original contributions presented in this study are included in the article and Supplementary Materials. Further inquiries can be directed to the corresponding author.

Acknowledgments: The authors would like to acknowledge the Research Institute for Renewable Energies of Politehnica University Timisoara for performing FE-SEM measurements and Politehnica University Timisoara, Romania, for the technical and financial support.

Conflicts of Interest: The authors declare no conflicts of interest.

Abbreviations

The following abbreviations are used in this manuscript:

C_{dl}	Double-layer capacitance
i_o	Exchange current density
R_{ct}	Charge transfer resistance
R_s	Solution resistance
AC	Alternating current
CP	Chronopotentiometry
CPE	Constant phase element
EDAX	Energy dispersive X-ray analysis
EEC	Equivalent electric circuits
EIS	Electrochemical impedance spectroscopy
FE-SEM	Field emission scanning electron microscopy
HER	Hydrogen evolution reaction
LSV	Linear sweep voltammetry
OER	Oxygen evolution reaction
PC	Pulsed current
PGE	Pencil graphite electrode
RHE	Reversible hydrogen electrode

References

1. McCrory, C.C.L.; Jung, S.; Ferrer, I.M.; Chatman, S.M.; Peters, J.C.; Jaramillo, T.F. Benchmarking Hydrogen Evolving Reaction and Oxygen Evolving Reaction Electrocatalysts for Solar Water Splitting Devices. *J. Am. Chem. Soc.* **2015**, *137*, 4347–4357. [CrossRef] [PubMed]
2. Ramakrishnan, S.; Vijayapradeep, S.; Selvaraj, S.C.; Huang, J.; Karthikeyan, S.; Gutru, R.; Logeshwaran, N.; Miyazaki, T.; Mamlouk, M.; Yoo, D.J. An efficient cathode electrocatalyst for anion exchange membrane water electrolyzer. *Carbon* **2024**, *220*, 118816. [CrossRef]
3. Gebremariam, G.K.; Jovanović, A.Z.; Pašti, I.A. Kinetics of Hydrogen Evolution Reaction on Monometallic Bulk Electrodes in Various Electrolytic Solutions. *Catalysts* **2023**, *13*, 1373. [CrossRef]
4. Zhu, Y.; Liu, T.; Li, L.; Song, S.; Ding, R. Nickel-based electrodes as catalysts for hydrogen evolution reaction in alkaline media. *Ionics* **2018**, *24*, 1121–1127. [CrossRef]
5. Harada, Y.; Hua, Q.; Harris, L.C.; Yoshida, T.; Gewirth, A.A. Electrodeposition of Fractal Structured Nickel for Hydrogen Evolution Reaction in Alkaline. *ChemElectroChem* **2024**, *11*, e202400005. [CrossRef]
6. Jawhari, A.H.; Hasan, N. Nanocomposite Electrocatalysts for Hydrogen Evolution Reactions (HERs) for Sustainable and Efficient Hydrogen Energy—Future Prospects. *Materials* **2023**, *16*, 3760. [CrossRef]
7. Cieluch, M.; Kazamer, N.; Böhm, L.; Sanden, S.; Zerebecki, S.; Wirkert, F.; Apfel, U.; Brodmann, M. Effect of Electrolyte pH in Additive-Free NiFe Catalyst Electrodeposition for Electro-Catalytic OER Applications. *ChemElectroChem* **2025**, *12*, e202400492. [CrossRef]
8. Raja, M.A.; Arumainathana, S. Comparative study of hydrogen evolution behavior of Nickel Cobalt and Nickel Cobalt Magnesium alloy film prepared by pulsed electrodeposition. *Vacuum* **2019**, *160*, 461–466. [CrossRef]
9. Feng, Y.; Guan, Y.; Zhou, E.; Zhang, X.; Wang, Y. Nanoscale Double-Heterojunctional Electrocatalyst for Hydrogen Evolution. *Adv. Sci.* **2022**, *9*, 2201339. [CrossRef]
10. Li, J.; Chu, D.; Poland, C.; Smith, C.; Nagelli, E.A.; Jaffett, V. XPS Depth Profiling of Surface Restructuring Responsible for Hydrogen Evolution Reaction Activity of Nickel Sulfides in Alkaline Electrolyte. *Materials* **2025**, *18*, 549. [CrossRef]
11. Bhat, K.S.; Nagaraja, H. Nickel selenide nanostructures as an electrocatalyst for hydrogen evolution reaction. *Int. J. Hydrogen Energy* **2018**, *43*, 19851–19863. [CrossRef]
12. Liu, G.; Hou, F.; Peng, S.; Wang, X.; Fang, B. Synthesis, Physical Properties and Electrocatalytic Performance of Nickel Phosphides for Hydrogen Evolution Reaction of Water Electrolysis. *Nanomaterials* **2022**, *12*, 2935. [CrossRef]
13. Danish, M.S.S. Exploring metal oxides for the hydrogen evolution reaction (HER) in the field of nanotechnology. *RSC Sustain.* **2023**, *1*, 2180. [CrossRef]
14. Zhu, Y.; Lin, Q.; Zhong, Y.; Tahini, H.A.; Shao, Z.; Wang, H. Metal Oxide-Based Materials as an Emerging Family of Hydrogen Evolution Electrocatalysts. *Energy Environ. Sci.* **2020**, *13*, 3361–3392. [CrossRef]

15. Huo, L.; Jin, C.; Jiang, K.; Bao, Q.; Hu, Z.; Chu, J. Applications of Nickel-Based Electrocatalysts for Hydrogen Evolution Reaction. *Adv. Energy Sustain. Res.* **2022**, *3*, 2100189. [CrossRef]
16. Sadhanala, H.K.; Perelshtein, I.; Gedanken, P.A. Boosting Electrocatalytic Hydrogen Evolution of Nickel foam Supported Nickel Hydroxide by Ruthenium Doping. *Chem. Sel.* **2020**, *5*, 9626–9634. [CrossRef]
17. Wang, H.; Xiong, J.; Cheng, X.; Fritz, M.; Ispas, A.; Bund, A.; Chen, G.; Wang, D.; Schaaf, P. Ni3N-Coated Ni Nanorod Arrays for Hydrogen and Oxygen Evolution in Electrochemical Water Splitting. *ACS Appl. Nano Mater.* **2020**, *3*, 10986–10995. [CrossRef]
18. Faid, A.Y.; Barnett, A.O.; Seland, F.; Sunde, S. Highly Active Nickel-Based Catalyst for Hydrogen Evolution in Anion Exchange Membrane Electrolysis. *Catalysts* **2018**, *8*, 614. [CrossRef]
19. Shen, Y.; Wu, P.; Wang, C.; Yuan, W.; Yang, W.; Shang, X. Electrodeposition of amorphous Ni-Fe-Mo composite as a binder-free and high-performance electrocatalyst for hydrogen generation from alkaline water electrolysis. *Int. J. Hydrogen Energy* **2023**, *48*, 33130–33138. [CrossRef]
20. Fernández-Valverde, S.; Ordóñez-Regil, E.; Cabañas-Moreno, J.G.; Solorza-Feria, O. Electrochemical Behavior of Ni-Mo Electrocatalyst for Water Electrolysis. *J. Mex. Chem. Soc.* **2019**, *54*, 169–174. [CrossRef]
21. Deng, Y.; Lai, W.; Xu, B. A Mini Review on Doped Nickel-Based Electrocatalysts for Hydrogen Evolution Reaction. *Energies* **2020**, *13*, 4651. [CrossRef]
22. Wang, X.; Tian, H.; Zhu, L.; Li, S.; Cui, X. Synergetic Catalytic Effect between Ni and Co in Bimetallic Phosphide Boosting Hydrogen Evolution Reaction. *Nanomaterials* **2024**, *14*, 853. [CrossRef] [PubMed]
23. Li, R.; Zang, P.J.; Li, P.W.; Li, J.; Zou, Q.; Zhou, S.; Su, J.; Wang, P.Y. Three-Dimensional Transition Metal Phosphide Heteronanorods for Efficient Overall Water Splitting. *ChemSusChem* **2020**, *13*, 3718–3725. [CrossRef]
24. Zhang, D.; Shi, J.; Qi, Y.; Wang, X.; Wang, H.; Li, M.; Liu, S.; Li, C. Quasi-Amorphous Metallic Nickel Nanopowder as an Efficient and Durable Electrocatalyst for Alkaline Hydrogen Evolution. *Adv. Sci.* **2018**, *5*, 1801216. [CrossRef] [PubMed]
25. Arcas, R.; Koshino, Y.; Mas-Marza, E.; Tsuji, R.; Masutani, H.; Miura-Fujiwara, E.; Haruyama, Y.; Nakashima, S.; Ito, S.; Fabregat-Santiago, F. Pencil graphite rods decorated with nickel and nickel–iron as low-cost oxygen evolution reaction electrodes. *Sustain. Energy Fuels* **2021**, *5*, 3929. [CrossRef]
26. Poimenidis, I.A.; Moustaizis, S.D.; Papakosta, N.; Tsanakas, M.D.; Klini, A.; Loukakos, P.A. Electrodeposition of Ni particles on laser nanostructured electrodes for enhanced hydrogen evolution reaction. *Mater. Today Proc.* **2022**, *67*, 953–958. [CrossRef]
27. Xu, C.; Chen, P.; Hu, B.; Xiang, Q.; Cen, Y.; Hu, B.; Liu, L.; Liu, Y.; Yu, D.; Chen, C. Porous nickel electrodes with controlled texture for the hydrogen evolution reaction and sodium borohydride electrooxidation. *CrystEngComm* **2020**, *22*, 4228–4237. [CrossRef]
28. Elsharkawy, S.; Kutyła, D.; Zabinski, P. The Influence of the Magnetic Field on Ni Thin Film Preparation by Electrodeposition Method and Its Electrocatalytic Activity towards Hydrogen Evolution Reaction. *Coatings* **2023**, *13*, 1816. [CrossRef]
29. Boubatra, M.; Azizi, A.; Schmerber, G.; Dinia, A. The influence of pH electrolyte on the electrochemical deposition and properties of nickel thin films. *Ionics* **2012**, *18*, 425–432. [CrossRef]
30. Wang, X.; Jiang, M.; Yang, P.; Zhou, H.; Xi, W.; Duan, J.; Ratova, M.; Wu, D.; Jiang, X. Recent advances in pulsed electrochemical techniques: Synthesis of electrode materials and electrocatalytic reactions. *Surf. Interfaces* **2024**, *50*, 104519. [CrossRef]
31. Boukhouiete, A.; Boumendjel, S.; Sobhi, N.E.H.; Creus, J. Microstructural investigation of nickel deposits obtained by pulsed current. *J. Ind. Chem. Soc.* **2022**, *99*, 100331. [CrossRef]
32. Balint, L.-C.; Hulka, I.; Kellenberger, A. Pencil Graphite Electrodes Decorated with Platinum Nanoparticles as Efficient Electrocatalysts for Hydrogen Evolution Reaction. *Materials* **2022**, *15*, 73. [CrossRef] [PubMed]
33. Torrinha, A.; Amorim, C.G.; Montenegro, M.C.; Araújo, A.N. Biosensing based on pencil graphite electrodes. *Talanta* **2018**, *190*, 235–247. [CrossRef] [PubMed]
34. Habibi, B.; Farhadi, K.; Minaie, E. Electrodeposited pectin/reduced carbon dots scaffold on the pencil graphite electrode as a support of electroloaded nickel nanoparticles for electrocatalytic purpose. *Int. J. Hydrogen Energy* **2024**, *91*, 1452–1462. [CrossRef]
35. Kayan, D.B.; Koçak, D.; İlhan, M.; Koca, A. Electrocatalytic hydrogen production on a modified pencil graphite electrode. *Int. J. Hydrogen Energy* **2017**, *42*, 2457–2463. [CrossRef]
36. Melo, L.C.; de Lima-Neto, P.; Correia, A.N. The influence of citrate and tartrate on the electrodeposition and surface morphology of Cu–Ni layers. *J. Appl. Electrochem.* **2011**, *41*, 415–422. [CrossRef]
37. Li, C.; Li, X.; Wang, Z.; Guo, H. Nickel electrodeposition from novel citrate bath. *Trans. Nonferrous Met. Soc. China* **2017**, *17*, 1300–1306. [CrossRef]
38. Bigos, A.; Wolowicz, M.; Janusz-Skuza, M.; Starowicz, Z.; Szczerba, M.J.; Bogucki, R.; Beltowska-Lehman, E. Citrate-based baths for electrodeposition of nanocrystalline nickel coatings with enhanced hardness. *J. Alloys Compd.* **2021**, *805*, 156857. [CrossRef]
39. Lasia, A.; Rami, A. Kinetics of hydrogen evolution on nickel electrodes. *J. Electroanal. Chem. Interfacial Electrochem.* **1990**, *294*, 123–141. [CrossRef]
40. Krstajič, N.; Popovič, M.; Grgur, B.; Vojnovič, M.; Sepa, D. On the kinetics of hydrogen evolution reaction on nickel in alkaline solution: Part I. The mechanism. *J. Electroanal. Chem.* **2001**, *512*, 16–26. [CrossRef]

41. Franceschini, E.A.; Lacconi, G.I.; Corti, H.R. Kinetics of the hydrogen evolution on nickel in alkaline solution: New insight from rotating disk electrode and impedance spectroscopy analysis. *Electrochim. Acta* **2015**, *159*, 210–218. [CrossRef]
42. Kellenberger, A.; Vaszilcsin, N.; Brandl, W.; Duteanu, N. Kinetics of hydrogen evolution reaction on skeleton nickel and nickel–titanium electrodes obtained by thermal arc spraying technique. *Int. J. Hydrogen Energy* **2007**, *32*, 3258–3265. [CrossRef]
43. Nikolic, V.M.; Maslovara, S.L.J.; Tasic, G.S.; Brdaric, T.P.; Lausevic, P.Z.; Radak, B.B.; Kaninski, M.P.M. Kinetics of hydrogen evolution reaction in alkaline electrolysis on a Ni cathode in the presence of Ni–Co–Mo based ionic activators. *Appl. Catal. B* **2015**, *179*, 88–94. [CrossRef]
44. Horvat-Radošević, V.; Kvastek, K. Impedance Spectra of the Hydrogen Evolution Reaction at Low Cathodic Overpotentials. *Croat. Chem. Acta* **2003**, *76*, 313–322.
45. Crețu, R.; Kellenberger, A.; Vaszilcsin, N. Enhancement of hydrogen evolution reaction on platinum cathode by proton carriers. *Int. J. Hydrogen Energy* **2013**, *38*, 11685–11694. [CrossRef]
46. Choquette, Y.; Brossard, L.; Lasia, A.; Menard, H. Study of the Kinetics of Hydrogen Evolution Reaction on Raney Nickel Composite-Coated Electrode by AC Impedance Technique. *J. Electrochem. Soc.* **1990**, *137*, 1723. [CrossRef]
47. Brug, G.J.; van den Eeden, A.L.G.; Sluyters-Rehbach, M.; Sluyters, J.H. The Analysis of Electrode Impedances Complicated by the Presence of a Constant Phase Element. *J. Electroanal. Chem. Interfacial Electrochem.* **1984**, *176*, 275–295. [CrossRef]
48. Gateman, S.M.; Gharbi, O.; Gomes de Melo, H.; Ngo, K.; Turmine, M.; Vivier, V. On the use of a constant phase element (CPE) in electrochemistry. *Curr. Opin. Electrochem.* **2022**, *36*, 101133. [CrossRef]
49. Perović, I.; Marčeta Kaninski, M.; Tasić, G.; Maslovara, S.; Laušević, P.; Seović, M.; Nikolić, V. Enhanced Catalytic Activity and Energy Savings with Ni-Zn-Mo Ionic Activators for Hydrogen Evolution in Alkaline Electrolysis. *Materials* **2023**, *16*, 5268. [CrossRef]
50. Manazoğlu, M.; Hapçı, G.; Orhan, G. Electrochemical Deposition and Characterization of Ni-Mo Alloys as Cathode for Al-kaline Water Electrolysis. *J. Mater. Eng. Perform.* **2016**, *25*, 130–137. [CrossRef]
51. Llorente, V.B.; Diaz, L.A.; Lacconi, G.I.; Abuin, G.C.; Franceschini, E.A. Effect of duty cycle on NiMo alloys prepared by pulsed electrodeposition for hydrogen evolution reaction. *J. Alloys Compd.* **2022**, *897*, 163161. [CrossRef]
52. Bao, F.; Kemppainen, E.; Dorbandt, I.; Bors, R.; Xi, F.; Schlatmann, R.; van de Krol, R.; Calnan, S. Understanding the Hydrogen Evolution Reaction Kinetics of Electrodeposited Nickel-Molybdenum in Acidic, Near-Neutral, and Alkaline Conditions. *ChemElectroChem* **2020**, *8*, 195–208. [CrossRef]
53. Tao, S.; Yang, F.; Schuch, J.; Jaegermann, W.; Kaiser, B. Electrodeposition of Nickel Nanoparticles for the Alkaline Hydrogen Evolution Reaction: Correlating Electrocatalytic Behavior and Chemical Composition. *ChemSusChem* **2018**, *11*, 948–958. [CrossRef] [PubMed]
54. Kuznetsov, A.N.; Oshchepkov, A.G.; Cherstiouk, O.V.; Simonov, P.A.; Nazmutdinov, R.R.; Savinova, E.R.; Bonnefont, A. Influence of the NaOH Concentration on the Hydrogen Electrode Reaction Kinetics of Ni and NiCu Electrodes. *ChemElectroChem* **2020**, *7*, 1438–1447. [CrossRef]
55. Solmaz, R.; Kardas, G. Electrochemical deposition and characterization of NiFe coatings as electrocatalytic materials for alkaline water electrolysis. *Electrochim. Acta* **2009**, *54*, 3726–3734. [CrossRef]

materials

MDPI

Article

Copper Slag Cathodes for Eco-Friendly Hydrogen Generation: Corrosion and Electrochemical Insights for Saline Water Splitting

Susana I. Leiva-Guajardo [1], Manuel Fuentes Maya [1,2,*], Luis Cáceres [3], Víctor M. Jimenez-Arevalo [4], Álvaro Soliz [5], Norman Toro [6], José Ángel Cobos Murcia [7], Victor E. Reyes Cruz [7], Mauricio Morel [8], Edward Fuentealba [1] and Felipe M. Galleguillos Madrid [1,*]

[1] Centro de Desarrollo Energético Antofagasta, Universidad de Antofagasta, Antofagasta 1240000, Chile; susana.leiva.guajardo@ua.cl (S.I.L.-G.); edward.fuentealba@uantof.cl (E.F.)
[2] Departamento de Ingeniería Mecánica, Universidad de Tarapacá, Arica 1100000, Chile
[3] Departamento de Ingeniería Química y Procesos de Minerales, Universidad de Antofagasta, Avenida Universidad de Antofagasta 02800, Antofagasta 1271155, Chile; luis.caceres@uantof.cl
[4] Departamento de Química de los Materiales, Facultad de Química y Biología, Universidad de Santiago de Chile, Av. Libertador B. O'Higgins 3363, Santiago 9170022, Chile; victor.jimenez@usach.cl
[5] Departamento de Ingeniería en Metalurgia, Universidad de Atacama, Av. Copayapú 485, Copiapó 1530000, Chile; alvaro.soliz@uda.cl
[6] Facultad de Ingeniería y Arquitectura, Universidad Arturo Prat, Iquique 1100000, Chile; notoro@unap.cl
[7] Instituto de Ciencias Básicas e Ingeniería, Universidad Autónoma del Estado de Hidalgo, Carr. Pachuca—Tulancingo, Mineral de la Reforma 42184, Hidalgo, Mexico; jose_cobos@uaeh.edu.mx (J.Á.C.M.); profe_4948@uaeh.edu.mx (V.E.R.C.)
[8] Departamento de Química y Biología, Universidad de Atacama, Av. Copayapú 485, Copiapó 1530000, Chile; mauricio.morel@uda.cl
* Correspondence: msfuentes@academicos.uta.cl (M.F.M.); felipe.galleguillos.madrid@uantof.cl (F.M.G.M.)

Abstract: The increasing demand for sustainable energy and clean water has prompted the exploration of alternative solutions to reduce reliance on fossil fuels. In this context, hydrogen production through water electrolysis powered by solar energy presents a promising pathway toward a zero-carbon footprint. This study investigates the potential of copper slag, an abundant industrial waste, as a low-cost electrocatalyst for the hydrogen evolution reaction (HER) in contact with saline water such as 0.5 M NaCl and seawater, comparing the electrochemical response when in contact with geothermal water from El Tatio (Atacama Desert). The physicochemical characterisation of copper slag was performed using XRD, Raman, and SEM-EDS to determine its surface properties. Electrochemical evaluations were conducted in 0.5 M NaCl and natural seawater using polarisation techniques to assess the corrosion behaviour and catalytic efficiency of the copper slag electrodes. The results indicate that copper slag exhibits high stability and promising HER kinetics, particularly in seawater, where its mesoporous structure facilitates efficient charge transfer processes. The key novelty of this manuscript lies in the direct revalorisation of untreated copper slag as a functional electrode for HER in real seawater and geothermal water, avoiding the use of expensive noble metals and aligning with circular economy principles. This innovative combination of recycled material and natural saline electrolyte enhances both the technical and economic viability of electrolysis, while reducing environmental impact and promoting green hydrogen production in coastal regions with high solar potential. This research contributes to the value of industrial waste, offering a viable pathway for advancing sustainable hydrogen technologies in real-world environments.

Keywords: copper slag; corrosion; physicochemical characteristic; electrochemical characteristic; sustainable hydrogen technologies

1. Introduction

Population growth and industrial activity drive an increasing demand for sustainable energy and clean water, leading to a greater dependence on fossil fuels. This reliance depletes the planet's finite resources and exacerbates environmental impacts [1]. The decarbonisation of the energy sector has underscored the need for alternative and clean solutions [2]. A promising approach is hydrogen production using solar energy, known as solar hydrogen (H_2). This is generated through water electrolysis powered by electricity from sunlight, making it sustainable, As solar energy is clean, abundant, and globally accessible. H_2 production is modular, scalable, and adaptable to varying demands. Its use represents an effective strategy for reducing CO_2 emissions [3–5].

Saline water electrolysis is a promising alternative for converting solar energy into hydrogen (H_2), contributing to the energy crisis. However, there is a crucial need to develop efficient, durable, and cost-effective electrocatalytic materials for the hydrogen evolution reaction (HER) [6]. Enhancing the design of these materials has become a key objective. In this context, copper slag, a by-product of the copper smelting process, is increasingly recognised as a valuable resource for emerging electrochemical technologies. Globally, approximately 30 million tonnes of copper slag are produced annually, with Chile contributing around 4.5 million tonnes [7]. Traditionally considered industrial waste, copper slag contains valuable metal oxides such as hematite (Fe_2O_3), covellite (CuS), and magnetite (Fe_3O_4), which exhibit synergistic electrocatalytic properties. Its abundance in copper-producing countries such as Chile makes it an attractive alternative to conventional materials [8–15].

An analysis of world metal consumption statistics reveals that copper ranks third globally, surpassed only by steel and aluminium. The primary geological genesis of copper, accounting for approximately 90% of global extraction, is associated with sulphide deposits (CuS-FeS) [16,17]. To obtain copper, the ore must be concentrated and separated from gangue through a high-temperature pyrometallurgical process. In this process, Cu_2S acts as a reducing agent, while oxygen from the air oxidises iron into iron oxide. This oxide then reacts with silica and lime, forming a silicate that is removed along with the slag. As a result, a significant amount of waste is generated, typically containing 30–40 wt% iron in oxide form, 25–40 wt% silica (mainly in the form of iron silicates), less than 11–12 wt% alumina and calcium oxides, 4–5 wt% other oxides (MgO, MnO, TiO_2, P_2O_5, Na_2O, K_2O), 1–1.5 wt% sulphur, and approximately 1–1.5 wt% of copper [17,18]. Studies suggest that copper slag can generate hydrogen at rates of up to 0.113 μmol/g·h under photocatalytic conditions [19]. The integration of seawater as an electrolyte in hydrogen production further enhances the sustainability of the process [20]. With an extensive coastline of 6435 km, Chile presents significant potential for hydrogen production in coastal areas with abundant solar energy. A prime example is the Atacama Desert (Chile), which stretches from Arica to La Serena, covering nearly 1600 km in length and up to 180 km in width. Known for its exceptionally high solar radiation levels, the Atacama Desert is a strategic region for green hydrogen projects. However, seawater contains dissolved salts and minerals that can affect hydrogen production efficiency and accelerate corrosion in critical electrochemical system components [21–23].

Despite these challenges, seawater has demonstrated potential as an electrolyte, particularly when combined with copper slag. Its mesoporous structure and high stability in saline environments allow it to resist corrosion while facilitating the electron transfer processes essential for hydrogen evolution. Several studies have explored hydrogen production using copper slag as a Fenton-type catalyst in advanced oxidation processes for water treatment. The presence of Fe^{2+} and Fe^{3+} oxidation states in copper slag enhances the generation of hydroxyl radicals, which play a crucial role in breaking down contaminants in wastewater [19,20,24]. The integration of copper slag into electrocatalytic systems for hydrogen production and water treatment addresses key environmental challenges while offering numerous sustainable benefits. Its use aligns with strategic objectives such as (i) waste valorisation, (ii) transforming industrial by-products into valuable resources, (iii) reducing slag accumulation, and (iv) generating added value from waste, thereby contributing to economic benefits. Moreover, copper slag-based electrocatalysis has the potential to significantly lower hydrogen production costs by eliminating reliance on expensive or rare semiconductor materials. Copper slag's electrocatalytic properties make it particularly well suited for the hydrogen evolution reaction (HER) in seawater. Research indicates that its mesoporous structure enhances the efficiency of seawater splitting for hydrogen production. Additionally, the interaction between copper slag and seawater promotes the generation of electron–hole pairs, which drive the oxygen reduction reaction (ORR), complementing HER and establishing a balanced redox process [19–21]. The use of such water sources offers several advantages, as seawater accounts for 96.5% of the Earth's total water resources and serves as a low-cost alternative. Furthermore, the presence of dissolved minerals such as chlorides, magnesium, and sodium can enhance the electrolysis process by increasing water's electrical conductivity. This, in turn, reduces the energy required to split water into hydrogen and oxygen, improving the overall process efficiency [25].

The novelty of this study lies in the revalorisation of copper slag as an electrocatalytic material for sustainable hydrogen production in saline water electrolysis as a novel application. This approach aligns with the principles of circular economy and material recycling, transforming industrial waste into a valuable resource for clean technologies. By using copper slag as a cathode in electrolysis processes, the need for expensive noble metals is eliminated, significantly reducing production costs. Moreover, the use of seawater as the electrolyte marks a crucial step towards sustainability, as it enables the utilisation of an abundant and low-cost resource, avoiding the use of freshwater in arid regions. This innovative combination of a recycled material and a natural saline medium not only improves the technical and economic viability of electrolysis but also reduces environmental impact and promotes the generation of green hydrogen in coastal areas with high solar potential. This study considered a qualitative characterisation using X-ray diffraction (XRD), Raman spectroscopy, and scanning electron microscopy with energy-dispersive spectroscopy (SEM-EDS), along with electrochemical evaluations in 0.5 M NaCl and, seawater, comparing the electrochemical response when the copper slags are in contact with geothermal water (El Tatio from Atacama Desert); the research assesses its structural, compositional, and catalytic properties. The findings demonstrate copper slag's high stability and efficiency in facilitating HER, particularly in saline environments, highlighting its potential as a cost-effective and sustainable material for green hydrogen production in coastal regions.

2. Materials and Methods

2.1. Copper Slag Preparation

Copper slag is a residue produced in a pyrometallurgical process to obtain copper concentrated from a concentrate of sulphide minerals, which contains materials such as FeO, SiO_2, Al_2O_3, CaO, and Cu [14]. In the pyrometallurgical stage known as smelting, very high-temperature furnaces are used. As a consequence of the high temperatures, immiscible liquid phases are produced in this process: the copper-rich matte (sulphide) and the copper slag called oxide [7]. The Antofagasta region of Northern Chile is among the main copper-producing regions of Chile. There are two copper smelters in the Antofagasta region. In Northern Chile, there are copper deposits of current economic relevance, such as Chuquicamata (Calama) and Alto Norte (Antofagasta), both of which are in an operational phase. In addition, the presence of remnants of old smelters, dating back to the 18th century, has generated an extensive dispersion of copper-bearing slag throughout the desert of this region. This accumulation of historical metallurgical by-products represents a potential source of geochemical and archaeometallurgical analysis [26]. The copper slag sample used for the preparation of the electrode is from a smelter in the Antofagasta region. Copper slag powder was obtained by comminuting the material to increase the available surface area for chemical reactions. Initially, copper slag rocks were crushed into smaller fragments to facilitate further processing. These fragments were then pulverised into finer particles, ensuring uniformity in particle size. Finally, the material was sieved to achieve a particle size of −400# mesh, resulting in a fine copper slag powder with optimal characteristics for subsequent applications.

2.2. Preparation of Copper Slag Electrodes

The electrode onto which the copper slag powder is deposited consists of a PTFE cylinder measuring 8 mm in diameter and 10 mm in length, serving as a fixing device on the shaft of the rotating disc electrode cell. A concave bronze rod (4 mm in diameter and 4 mm in length) is inserted concentrically into this cylinder. In preparing the working electrode, a modified carbon paste was formulated by homogenising graphite powder (50% w/w) and solid paraffin (50% w/w). Modified carbon paste is used to prepare copper slag electrodes because of its wide availability, good conductivity, and ease of preparation [27]. This resulting paste was carefully deposited on a bronze rod with a concave configuration, designed to ensure efficient electrical conductivity with the rotating disc electrode (RDE) system. A quantity of copper slag powder is then applied and pressed to adhere to the surface. The assembled electrode was integrated into the RDE equipment for electrochemical analysis. To ensure the robustness and representativeness of the electrochemical data, the complete preparation and measurement procedure was replicated three times (see Figure 1).

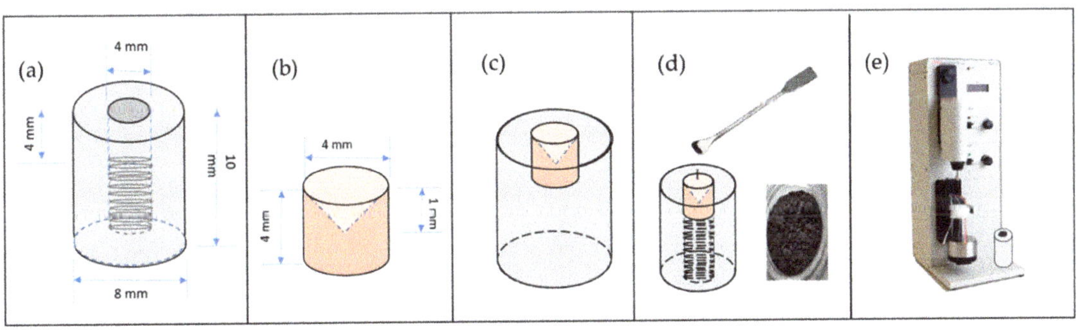

Figure 1. Construction of copper slag electrode. (**a**) PTFE cylinder, (**b**) concave bronze rod, (**c**) bronze rod inserted into PTFE cylinder, (**d**) copper slag powder paste deposited onto rod, (**e**) assembled electrode was integrated into RDE equipment.

2.3. Preparation of Saline Electrolyte

Saline solutions of 0.5 M NaCl were prepared using analytical-grade with ≥99.5% of purity of NaCl from Merck (Merck, Darmstadt, Germany) and deionised water. Seawater and geothermal water (El Tatio) were also used due to their high contents of ions such as chlorides, sodium, and magnesium, which act as natural electrolytes. These ions enhance the conductivity of the medium, making seawater suitable for electrochemical studies to evaluate charge transfer processes and redox reactions under conditions similar to real environmental scenarios [25]. Zhao et al. presented the chemical composition of seawater across different marine regions globally. Their research underscores that seawater primarily consists of dissolved ions, significantly determining its salinity and physicochemical properties. The average concentrations of the principal ions in seawater, measured in milligrams per litre (mg/L), are as follows: Cl^- (19.229 mg/L), Na^+ (1.114 mg/L), Mg^{2+} (1.279 mg/L), Ca^{2+} (359 mg/L), K^+ (340 mg/L), and $SO_4{}^{2-}$ (2.649 mg/L). El Tatio geothermal water plentiful in hot springs and geysers, located in the Andes in the region of Antofagasta, Northern Chile, 80 km from San Pedro de Atacama (22°20′ S, 68°01′ W), at an altitude of 4200 m a.s.l. It has a pH of 6 to 8 and a conductivity of ~20 mS/cm. Its discharged waters have high concentrations. The geothermal water (El Tatio) composition is as follows: Cl^- (6000–8000 mg/L), Na^+ (>3500 mg/L), $SO_4{}^{2-}$ (<50 mg/L), Mg^{2+} (50–150 mg/L), Ca^{2+} (100–300 mg/L), and $HCO_3{}^-$ (200–500 mg/L) [28,29]. These values represent the general chemical composition of seawater worldwide, though minor variations can be observed depending on the oceanic region, depth, and specific environmental conditions [30,31]. Among these constituents, Cl^- and Na^+ are the predominant electrolytes, playing a fundamental role in ion transport and enhancing the electrical conductivity of seawater.

2.4. *Characterisation of Copper Slag Powder*

2.4.1. X-Ray Diffraction (XRD)

The XRD patterns of the copper slag powders were recorded on a Bruker Advance D8 diffractometer (Billerica, MA, USA). A Cu Kα1 (λ = 1.5406 Å) radiation source a Vertical Bragg-Brentano goniometer set at 40 kV and 30 mA and a scintillation detector were used. The XRD patterns were obtained in the 2θ range from 10° to 70°, with a step size and dwell time of 0.020° (2θ) and 3.0 s per step, respectively.

2.4.2. Scanning Electron Microscopy (SEM-EDS)

General characterisation was performed using scanning electron microscopy (SEM) to reveal the morphological features of the microstructure. Surface analysis was carried out by SEM coupled with an energy-dispersive X-ray (EDX) spectrometer, employing a Zeiss EVA MA 10 microscope (Zeiss, Oberkochen, Germany). An accelerating voltage ranging from 5 kV to 15 kV was applied.

2.4.3. Raman Spectroscopy

Raman spectra were obtained using a Raman Confocal Microscope, LabRAM Soleil, serial number: 1068, equipped with a 532 nm laser (gratings: 600 lines/nm) Horiba (Chelmsford, Essex, UK). LabSpec 6 and OriginPro 2023 (64-bit) (Copyright © 1991-2022 OriginLab Corporation, Northampton, MA, USA), were used to analyse the spectra.

2.5. *Electrochemical Studies of Copper Slag Powder*

Electrochemical analyses were carried out using a BASi RDE-2 rotating electrode interface connected to an Epsilon potentiostat/galvanostat (Basi, West Lafayette, IN, USA). The analyses were performed separately by linear sweep voltammetry (LSV) in a conventional three-electrode cell setup, using copper slag powder as the working electrode, a platinum (Pt) wire as the counter electrode, and a silver/silver chloride (Ag/AgCl (4 M KCl sat.)) electrode as the reference. The experimental protocol for polarisation data was as follows: The cell was filled with the saline electrolyte and bubbled with air for approximately 15 min until an oxygen-saturated concentration was reached. The working electrode was then inserted into the cell, and linear sweep voltammetry was conducted within a potential range from −1200 to +200 mV/SHE at a scan rate of 2 mV/s. The rotation speed for the copper slag electrode was set to 1200 rpm.

2.6. Corrosion Studies for Copper Slag Powder

The experimental corrosion procedure was designed to examine the corrosion behaviour of partial electrochemical reactions on copper slag powder electrodes immersed in 0.5 M NaCl, seawater, and geothermal (El Tatio) solutions, focusing on the hydrogen evolution reaction (HER), oxygen reduction reaction (ORR), and copper slag oxidation reaction (CSOR). The temperature was maintained at 20 ± 0.5 °C using a water-jacketed cell that circulated water through a thermoelectric temperature control device. All experiments were repeated in triplicate. All potential measurements are referenced to the standard hydrogen electrode (SHE). The kinetic corrosion investigation was conducted using a non-linear fitting that was applied to experimental polarisation data [32]. The superposition model based on mixed potential theory was employed, incorporating both cathodic and anodic processes. The analysis focused on charge transfer, mass diffusion, and passivation mechanism controls, where the total current density can be expressed in Equation (1).

$$i = i_{O_2} + i_{H_2} + i_{CS} \tag{1}$$

where i represents the total current density, i_{O_2}, i_{H_2}, and i_{CS} represent the partial reduction current densities for the oxygen reduction reaction (ORR) and hydrogen evolution reaction (HER), respectively, and i_{CS} represents the partial oxidation current density for the copper slag oxidation reaction (CSOR). Although the experimental measurement of i_{O_2}, i_{H_2}, and i_{CS} is not directly feasible, their values can be deduced by considering the kinetic expressions for each reaction.

For this purpose, the partial reactions for HER and CSOR are modelled considering a charge transfer kinetic mechanism, while the ORR is described using a mixed charge-diffusion kinetic mechanism, according to the following expressions:

$$i_{H_2} = i_{0,H_2} exp\left(-\frac{2.303 \cdot \eta_{H_2}}{t_{H_2}}\right) \tag{2}$$

$$i_{O_2} = i_{0,O_2} exp\left(-\frac{2.303 \cdot (E - E_{eq,\,O_2})}{t_{O_2}}\right) \cdot \left(1 - \frac{i_{O_2}}{i_{l,O_2}}\right)^{0.5} \tag{3}$$

$$i_{CS} = i_{0,CS} exp\left(\frac{2.303 \cdot \eta_{CS}}{t_{CS}}\right) \tag{4}$$

where i_{0,H_2}, i_{0,O_2}, $i_{0,CS}$, are the exchange current densities for HER, ORR, and CSOR, respectively. i_{l,O_2} represents the limiting current density for ORR. The overpotentials for HER, ORR, and CSOR are expressed as son $\eta_{H_2} = E - E_{eq\ H_2}$, $\eta_{O_2} = E - E_{eq\ O_2}$ and $\eta_{CS} = E - E_{eq\ CS}$, respectively. The equilibrium potentials for HER, ORR, and CSOR are denoted as $E_{eq\ H_2}$, $E_{eq\ O_2}$, and $E_{eq\ CS}$, respectively. The cathodic Tafel slopes for HER and ORR are t_{H_2} and t_{O_2}, respectively, while t_{CS} represents the anodic Tafel slope for CSOR.

The exchange current densities $(i_{o,\,j})$, have been determined using the Butler–Volmer equation at the equilibrium potential for each partial reaction involved in the electroanalysis. To achieve this, we have used electrochemical parameters obtained from the non-linear fitting of experimental data. The equation can be simplified as follows:

$$i_{o,j} = a\ exp\left(b \cdot E_{eq,j}\right) \tag{5}$$

where a is base current (i) and b is sensitivity to potential, are the fitted parameters derived from Equation (1).

3. Results

Structural and Morphological Analysis of Copper Slag

Figure 2 shows the X-ray diffraction (XRD) spectra. The figure illustrates the copper slag dust patterns, showing that the main phases present in this material are fayalite (Fe_2SiO_4) and magnetite (Fe_3O_4).

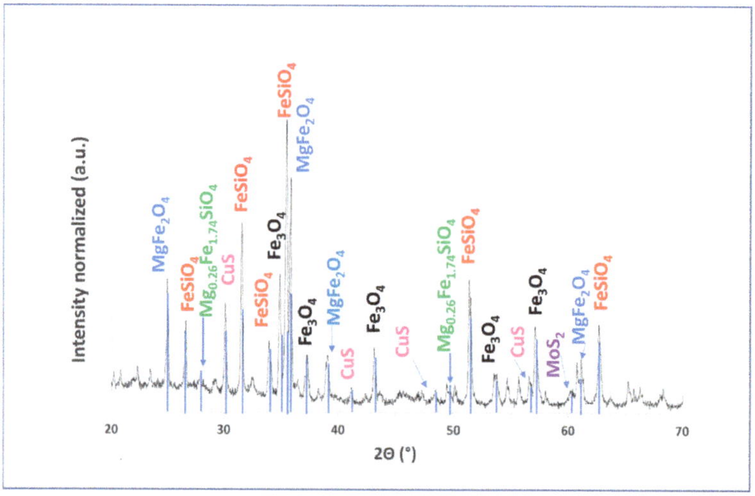

Figure 2. XRD diffractogram of copper slag powder showing peaks for fayalite and magnesioferrite. Data were collected using Bruker Advance D8 diffractometer (Cu Kα1, λ = 1.5406 Å) with 2θ range of 10–70°, 0.020° step size, and 3.0 s dwell time per step.

Figure 3 shows an image of copper slag powder before contact with salt solutions through the scanning electron microscope (SEM), the signal is under BSD1, magnitude 1.74 KX, ETH 5.00 Kv, WD 7.2 mm, and at 10 µm. It was observed that the copper slag powder shows a diversity of crystalline phases. Mostly angular and flaky particles are observed. In addition, the copper slag powder shows the presence of elements of higher molar mass due to its brightness.

Figure 3. SEM image of copper slag powder before exposure to saline solutions, acquired under BSD1 mode at 1.74 KX magnification, 5.00 kV accelerating voltage, 7.2 mm working distance, and 10 µm scale.

The point analysis, presented in Figure 4, reveals, in general terms, that iron is present in 12.1 wt%, oxygen represents 28.2 wt%, and silicon accounts for 7.3 wt%. This suggests that the copper slag powders have a high iron oxide content.

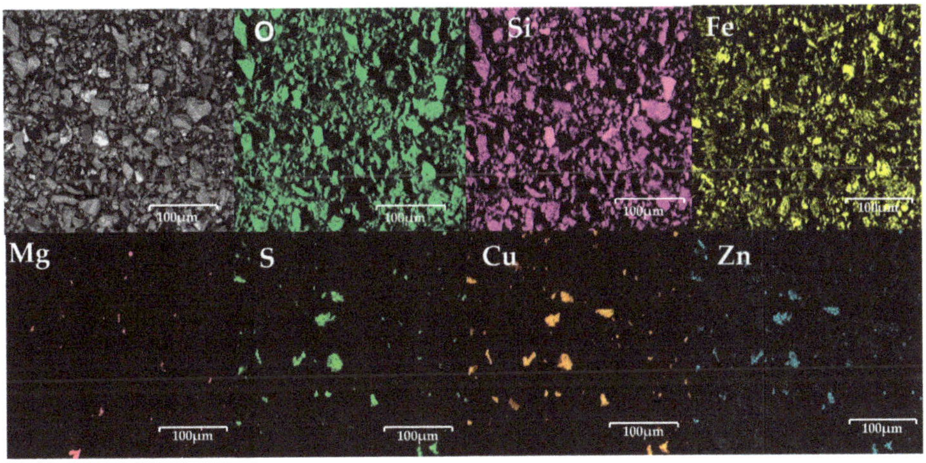

Figure 4. Point analysis of copper slag powder electrodes before contact with seawater, showing distribution of key elements such as oxygen, silicon, iron, and magnesium, as well as presence of residual sulphur, copper, and zinc.

The Raman spectrum analysis shows the presence of magnetite (Fe_3O_4), hematite (Fe_2O_3), and olivine (fayalite type $(Mg,Fe)_2SiO_4$). In Figure 5a, the Raman spectrum of a darker zone of the sample is shown, which coincides with the main presence of magnetite. The analysis of the Raman spectrum of the copper slag presented in Figure 5b, which shows the spectrum of a lighter area, presents mostly fayalite in the form of olivine.

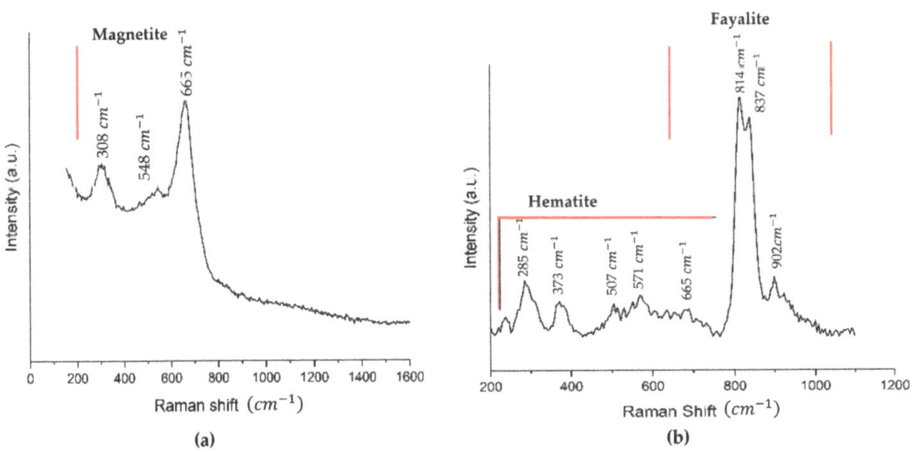

Figure 5. (**a**) The Raman spectra of the dark areas of the copper slag sample, (**b**) In the lightest part of the sample, the olivine-type fayalite can be seen. The red line indicates the zone where are representative signal of each component (magnetite, hematite and fayalite).

Figure 6 shows the copper slag modified carbon paste electrode after the oxidation process in 0.5 M NaCl. The results revealed the presence of cubic crystals on the electrode.

Elemental analysis via EDS, as shown in Figure 6, reveals the presence of cubic crystals of NaCl, indicating the contribution of the NaCl solution to the observed composition.

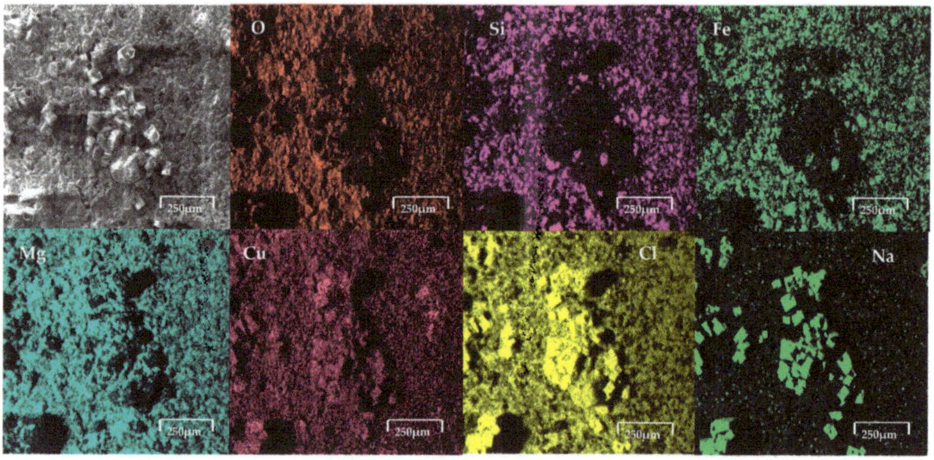

Figure 6. SEM-EDS analysis of copper slag powder electrodes after contact with saline. It clearly shows presence of cubic crystals on electrode surface. These crystals, according to EDS, correspond to NaCl.

Figure 7 shows the modified carbon paste electrode with copper slag after the oxidation process in seawater. The EDS analysis confirmed the presence of elements characteristic of copper slag powder, including Cu (5.2 wt%), Fe (39.9 wt%), Si (13.4 wt%), and O (29.2 wt%), along with the incorporation of ions from the saline medium, such as ions Cl^- and Na^+, respectively.

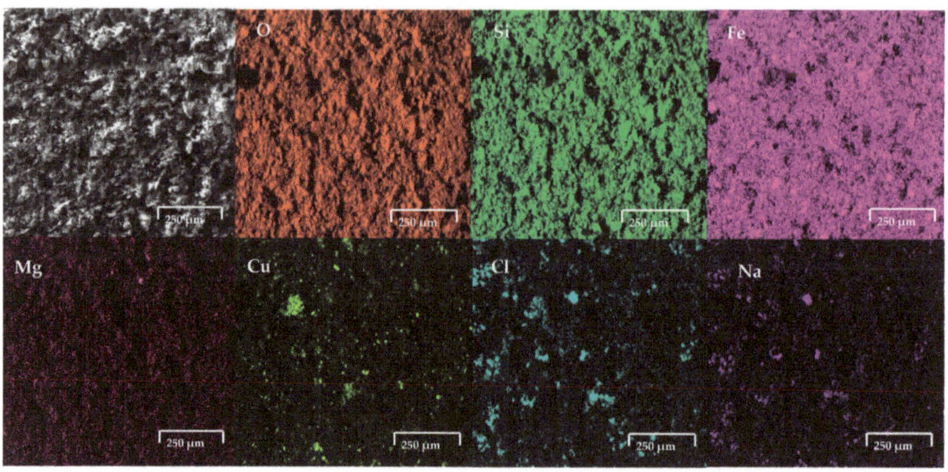

Figure 7. SEM-EDS analysis of copper slag powder electrodes after contact with seawater.

Figure 8 provides important insights into the electrochemical kinetic performance of copper slag powder during the sub-cathodic process in contact with 0.5 M NaCl, seawater, and geothermal water from El Tatio (Atacama Desert), respectively.

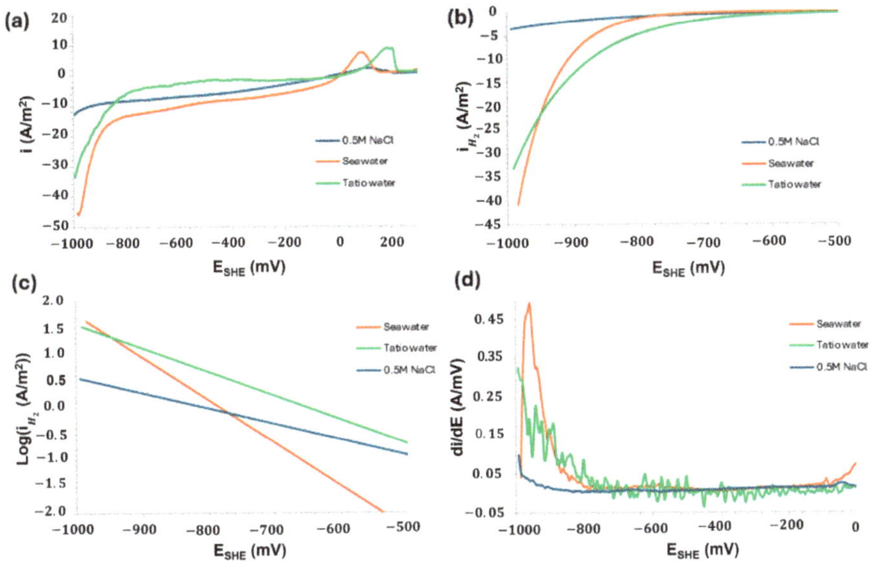

Figure 8. Electrochemical performance of 0.5 M NaCl, seawater, and geothermal water (El Tatio, Atacama Desert). (**a**) Lineal voltammetry (LSV), (**b**) H_2 evolution, (**c**) H_2 Tafel slope, and (**d**) behaviour of first derivative (di/dE).

4. Discussion

Previous studies, such as that by Allibai et al. [33], focus on the development of electrochemical sensors based on nanocomposites like Fe_2O_3 and copper hexacyanoferrate (CuHCF) for detecting specific analytes such as hydroxylamine in aqueous media. This manuscript proposes a broader and more impactful application in the context of the energy transition: the direct use of copper slag as an electrocatalytic material for the hydrogen evolution reaction (HER) in seawater electrolysis. In contrast, the work of Li et al. [34]. centres on the chemical and mineralogical characterisation of copper slag, evaluating its environmental hazards through leaching and sequential extraction tests, with an emphasis on the recovery of metals such as iron, copper, and lead. The current study goes a step further by actively valorising copper slag through its direct integration into electrolysis cells, without requiring prior mineral separation or concentration processes. This strategy not only reduces costs and material waste but also aligns with the principles of circular economy and water sustainability—particularly significant in one of the driest regions on Earth. Taken together, this research offers an innovative, multifunctional, and environmentally responsible approach the utilisation of industrial waste, paving the way for the sustainable and decentralised production of green hydrogen [33,34].

Iron oxides and sulphides—such as magnetite (Fe_3O_4), haematite (Fe_2O_3), and pyrite (FeS_2)—have emerged as promising electrocatalytic materials owing to their low toxicity, high abundance, low cost, and favourable chemical stability. These compounds possess semiconducting properties and well-defined crystalline structures that facilitate charge transfer and the adsorption of reactive species during electrochemical processes such as the hydrogen evolution reaction (HER). Specifically, magnetite and haematite can serve as active sites for water molecule dissociation, thereby promoting the generation of molecular hydrogen. Pyrite and other iron sulphides have also demonstrated notable electrocatalytic behaviour, particularly when incorporated into hybrid materials or supported on conductive substrates. Recent studies have shown that the modification of electrodes with Fe_2O_3

nanoparticles significantly enhances HER performance by increasing the electroactive surface area and improving charge transfer kinetics. Similarly, synergistic effects arising from combinations of metal sulphides and mixed oxides have been shown to enhance catalytic activity, even under saline conditions. Given their natural abundance and compatibility with waste-derived systems such as copper slag, these materials are well-positioned to contribute to the development of green hydrogen technologies within the framework of sustainable energy transition [20,33,35].

Research into pyrite (FeS_2) and its derivatives as electrocatalysts has garnered growing attention due to their natural abundance, low cost, and advantageous structural characteristics for key electrochemical processes such as the oxygen evolution reaction (OER) and hydrogen evolution reaction (HER). Studies have shown that natural pyrite, when supported on conductive substrates such as nickel foam, exhibits exceptionally high catalytic activity, often comparable to that of IrO_2, highlighting the critical role of synergistic interactions between the catalyst and its support. In parallel, various strategies have been explored to enhance catalytic efficiency and address limitations such as weak proton (H^+) adsorption commonly found in certain iron sulphides. These strategies include heteroatomic alloying (e.g., Ni doping in FeS_2), nanostructuring, and fine-tuning of crystallographic parameters such as crystallite size and lattice strain. Such modifications have been shown to significantly improve the bifunctional activity of pyrite in alkaline media, with performance approaching that of commercial catalysts like RuO_2–Pt/C. Beyond OER and HER, recent studies have revealed the potential of FeS_2 as a photocatalyst for electrochemical nitrogen reduction, further underscoring its functional versatility. Structural variants such as cobalt pyrite (CoS_2) and hybrid systems like CoPS supported on carbon nanotubes have also demonstrated low overpotentials and high current densities. These findings emphasise that controlled synthesis, targeted elemental substitution (e.g., P or Ni doping), and integration with conductive frameworks are key strategies for enhancing electrocatalytic performance. Taken together, these advances reinforce the viability of pyrite and its derivatives as sustainable and efficient alternatives to noble metal-based catalysts, consolidating their relevance in the transition towards clean, affordable, and decentralised energy conversion technologies [36–42].

During the cathodic subprocess of electrolysis in seawater, in addition to the hydrogen evolution reaction (HER), critical operational issues may arise that affect the system's efficiency and durability. One of the most significant is the formation of inorganic scaling, caused by the precipitation of poorly soluble salts such as calcium carbonate ($CaCO_3$), magnesium hydroxide ($Mg(OH)_2$), and calcium sulphate ($CaSO_4$), induced by the local increase in pH at the cathode interface. These deposits reduce the active surface area of the electrode, increase charge transfer resistance, and may obstruct gas release channels, thereby lowering the efficiency of hydrogen production. In parallel, seawater contains a significant load of microorganisms, nutrients, and organic matter, which promote the formation of biofouling layers on the cathode surface [43–45]. These biological films can act as diffusional barriers, alter the surface microstructure of the electrode, and facilitate localised corrosion, particularly in materials with chemical heterogeneity such as copper slags. The simultaneous presence of mineral scaling and biofouling presents major challenges for the continuous and stable operation of the system [46]. Therefore, assessing the resistance of residual materials such as copper slag to these phenomena under real conditions (i.e., untreated seawater) is essential to determine their practical viability as cathodes in low-cost and sustainable electrolysis technologies.

It is true that materials such as platinum (Pt) exhibit exceptional catalytic activity and are the benchmark for the hydrogen evolution reaction (HER) under ideal conditions [47,48]. However, the aim of this study is not to compete directly with noble metals or advanced materials synthesised under controlled laboratory conditions, but rather to evaluate the potential of copper slag as a low-cost and functional electrocatalyst under real-world conditions, using untreated seawater as the electrolyte. From this perspective, although the current density achieved by copper slag in 0.5 M NaCl is lower than that reported for commercial catalysts, its electrochemical stability and activity in complex media such as seawater and geothermal water (El Tatio) represent a significant step forward towards practical, sustainable, and low-cost applications, particularly in arid regions such as Northern Chile, where this type of waste is readily available and access to freshwater is limited.

Moreover, there is growing interest in the use of natural or waste-derived materials as electrodes. For example, the use of natural pyrite (FeS_2) has been reported with promising results for OER on nickel foam, and strategies such as doping and nanostructuring have been explored to improve the performance of metal sulphides like CoS_2 and MoS_2, though with greater synthetic complexity and associated costs [49]. In this context, copper slag offers a direct alternative without the need for prior modification processes, making it easier to scale up and integrate into distributed hydrogen generation systems. We acknowledge the importance of contextualising our results in comparison with other electrocatalysts. In future versions of the manuscript, a comparative table will be included featuring representative materials such as Pt/C, MoS_2, CoP, and mixed oxides, considering key parameters such as overpotential at 10 mA cm^{-2}, Tafel slope, stability, and cost per unit of active area. This comparison will help to position copper slag more clearly as a viable alternative in decentralised and environmentally responsible electrolysis technologies [34,50–53].

Figure 2 illustrates patterns of copper slag dust, showing that the main phases present in this material are fayalite (Fe_2SiO_4) and magnetite (Fe_3O_4). According to Li et al., it is possible to identify complex silicates and other residual metallic elements in addition to these main phases [34]. Additionally, other phases observed in the copper slag dust sample include magnesioferrite ($MgFe_2O_4$), molybdenite (MoS_2), and, to a lesser extent, the crystalline phase of covellite (CuS). Furthermore, taking into account the known origin of the sample, it can be stated that amorphous phases corresponding to mineraloids were identified. Furthermore, previous studies suggest that XRD may not detect certain residual metallic elements due to their low concentration or because they are dispersed within the copper slag [9,26,34,35,54,55].

The point analysis, presented in Figure 4, reveals, in general terms, that iron is present in 12.1 wt%, oxygen represents 28.2 wt%, and silicon accounts for 7.3 wt%. This suggests that the copper slag powders have a high iron oxide content. Additionally, 4.0% and 0.5 wt% were detected, indicating that a small amount of copper sulphide still remains in the slag powders. Given this composition, the copper slag exhibits significant potential for electrochemical applications, particularly due to its iron oxide content. In this regard, Allibai et al. demonstrated that modifying a wax-impregnated graphite electrode with iron oxide (Fe_2O_3) nanoparticles enhances the electron transfer kinetics and electrocatalytic activity while increasing the surface area of the electrode. This results in improved anodic responses for electrochemical sensing, specifically for hydroxylamine detection [33]. Considering the high concentration of iron oxide in the copper slag, it may be a promising candidate for similar electrochemical applications, where such enhancements could be beneficial.

The Raman spectrum analysis shows the presence of magnetite (Fe_3O_4), hematite (Fe_2O_3), olivine ($(Mg, Fe)_2SiO_4$), and fayalite (Fe_2SiO_4). In Figure 5a, the Raman spectrum of a darker zone of the sample is shown, which coincides with the main presence of magnetite. This reported band indicates three signals, two weak bands at ~308 1/cm (Eg) and ~548 1/cm (T2g) and one prominent band at ~665 1/cm. Das et al. indicate that the prominent bands of magnetite, whose broadband range for magnetite is ~661 to ~676 1/cm (A1g), are consistent with other authors [56]. The analysis of the Raman spectrum of the copper slag presented in Figure 5b shows the spectrum of a lighter area, which presents mostly fayalite in the form of olivine, showing three characteristic signals at ~813, ~837, and ~902 1/cm, respectively. Kuebler et al. observed that the positions of the olivine doublet peaks vary with composition, from olivine to forsterite [57]. Olivine has a characteristic signal as a doublet around 837–815 1/cm for symmetric Si-O stretching band and has one symmetric Si-O stretching band between 825 and 808 1/cm. According to the XRD pattern, this corresponds mainly to fayalite with some complement of forsterite with a composition containing magnesium (Mg) below 0.26 and iron (Fe) above 1.74. Hematite may have appeared by oxidation of magnetite upon interaction with the laser [58,59]. This indicates that the sample analysed belongs to the olivine solid solution series, with a predominance of its iron-rich phase [60].

The electrodes, composed of copper slag powder and carbon wax, were immersed separately in seawater and an aqueous solution with a high salt concentration and subjected to electrochemical analysis to assess their response to corrosion and other electrochemical processes. Following this procedure, the electrodes were dried and prepared for characterisation using SEM-EDS. The SEM technique was employed to investigate morphological alterations on the material surface, while EDS analysis enabled the determination of its surface chemical composition. The images obtained by SEM revealed modifications in the surface morphology of the electrode, characterised by the presence of a porous structure and crystalline formations on the surface. This suggests the occurrence of selective dissolution processes and the precipitation of compounds within the saline medium. Figure 6 shows the copper slag modified carbon paste electrode after the oxidation process in 0.5 M NaCl. The results revealed the presence of cubic crystals on the electrode. From EDS analysis, as presented in the figure, the elemental analysis indicates that the formation of cubic crystals is NaCl, which comes from the NaCl solution. This indicates that there may be NaCl crystals and ions present in seawater (no presence of $CaCO_3$, $Mg(OH)_2$ and $CaSO_4$ was detected). This suggests that the copper slag dusts absorbed salts on their surface. Figure 7 shows the modified carbon paste electrode with copper slag after the oxidation process in seawater. The EDS analysis confirmed the presence of elements characteristic of copper slag powder, including Cu (5.2 wt%), Fe (39.9 wt%), Si (13.4 wt%), and O (29.2 wt%), along with the incorporation of ions from the saline medium, such as Cl^- and Na^+. The detection of these elements indicates the formation of corrosion products and the deposition of salts on the electrode surface. The chemical composition determined by EDS following exposure to seawater was 2.9 wt% Na, 3.3 wt% Al, 1.5 wt% Cl, 1.2 wt% Ca, 1.4 wt% K, 0.7 wt% Mg, and 2.3 wt% Mo.

The kinetic study was performed by applying a non-linear fit to the experimental polarisation data, based on the superposition model and mixed potential theory. This theoretical framework assumes that both anodic and cathodic reactions are governed by charge transfer, mass diffusion, and the control of oxidation mechanisms. Figure 8 provides key insights into the electrochemical and kinetic behaviour of the copper slag powder electrode during the cathodic subprocess, as a function of the electrolyte used (0.5 M NaCl, seawater, and El Tatio geothermal water). Figure 8a presents the polarisation curves, revealing a clear cathodic response. Among the tested electrolytes, seawater exhibited the highest cathodic current, followed by geothermal water and, lastly, 0.5 M NaCl. This trend suggests that natural electrolytes possess higher conductivity, likely due to their complex ionic composition and enhanced oxygen availability—factors that favour the oxygen reduction reaction (ORR). These findings are consistent with those reported by Santos et al. and Tong et al., who observed improved cathodic activity in natural saline media, attributed to their greater ionic complexity [61,62]. The negative potential region (Figure 8b) enables a kinetic-focused analysis. In this region, seawater exhibits a steeper slope, indicating a higher exchange current density within the same potential range. The distinct Tafel slopes observed for the three electrolytes (Figure 8c) suggest that the HER mechanism on the copper slag surface is influenced by the specific composition of each electrolyte. Notably, the Tafel slope for 0.5 M NaCl is higher than those for seawater and El Tatio geothermal water, implying slower charge transfer kinetics in the NaCl solution. The differential derivative analysis (di/dE, Figure 8d) reveals abrupt changes in the reaction mechanisms. Peaks observed in seawater and geothermal water around −950 mV may be attributed to redox-active components within the electrolyte. In contrast, the curve for 0.5 M NaCl exhibits a more uniform profile, reflecting a more stable but less electrochemically active system.

The specific behaviour around 0 V/SHE, corresponding to seawater, is not as pronounced in 0.5 M NaCl. It can be attributed to the transition of the ORR mechanism, dissolution of passive layers, or the reduction in residual species. Due to the characteristic of the electropotential, which contains chloride ions and is in materials such as copper slag, which could have traces of Cu or metal sulphides that are reduced around 0 V vs. SHE, generating these additional anodic peaks. Grimm et al. demonstrated this effect in copper materials using saline electrolytes [63].

The comparative electrochemical and kinetic parameters of copper slag powder are summarised in Table 1. Copper slag exhibits greater activity for both the oxygen reduction reaction (ORR) and the hydrogen evolution reaction (HER) in seawater. According to Kaiser et al., copper alloys in saline environments display reduced corrosion resistance, which may contribute to the observed behaviour. The highest exchange current density was recorded in geothermal water, suggesting that the heterogeneous surface of the copper slag enhances its catalytic activity for the ORR under these conditions [64]. Studies by Yaro et al. and Khadom et al. indicate that corrosion kinetics are influenced by both the chemical composition of the electrolyte and the nature of the electrode material. The corrosion analysis further revealed that, although the corrosion potential (E_{corr}) shifted towards more positive values and the corrosion current density (i_{corr}) increased in seawater, this electrolyte's composition accelerates corrosion in copper slag [65,66]. In contrast, Metikos-Hokovic et al. emphasise that both flow dynamics and surface composition affect passive film formation, which supports the lower corrosion rate observed in geothermal water, evidenced by its more positive (E_{corr}) and reduced (i_{corr}) values [67].

Table 1. Electrochemical parameters and kinetic properties of copper slags in contact with salt solutions.

Parameters	0.5 M NaCl	Seawater	Geothermal Water (El Tatio)
$i_{0,CS}$, A m^{-2}	0.1066	0.0021	0.0011
t_{CS}, mV dec^{-1}	338	140	157
i_{0,O_2}, A m^{-2}	0.097	0.380	0.0186
t_{O_2}, mV dec^{-1}	−587	−764	−417
i_{l,O_2}, A m^{-2}	−7.8	−11.9	−2.1
i_{0,H_2}, A m^{-2}	−0.383	−84.896	−0.132
t_{H_2}, mV dec^{-1}	−263	−126	−221.4
E_{corr}, mV	−1.124	12.539	30
i_{corr}, A m^{-2}	2.117	3.668	1.02

5. Conclusions

This study demonstrates the feasibility of using untreated copper slag as a low-cost and sustainable electrocatalyst for hydrogen production through saline water electrolysis.

Through comprehensive physicochemical and electrochemical characterisation—including XRD, SEM-EDS, Raman spectroscopy, and kinetic modelling—we establish that copper slag exhibits favourable surface properties, high electrochemical stability, and catalytic activity toward the hydrogen evolution reaction (HER), particularly in seawater and geothermal brine environments. These findings underscore the potential of this abundant industrial by-product as a functional electrode material that bypasses the need for expensive noble metals.

Notably, the material's performance in real electrolytes such as seawater and El Tatio geothermal water validates its robustness under complex, high-salinity conditions. The electrochemical response confirms that copper slag not only withstands corrosive media but also facilitates effective charge transfer, positioning it as a viable option for decentralised, low-cost hydrogen generation systems in arid coastal regions like Northern Chile.

This approach aligns with circular economy principles, offering a dual benefit: waste valorisation and green hydrogen production. Future work should focus on the long-term operational stability of copper slag electrodes under continuous use, the mitigation of scaling and biofouling effects, and the integration of this material into pilot-scale electrolysis systems. Such efforts will be essential to assess its commercial scalability and full environmental impact in sustainable energy applications.

Author Contributions: Conceptualization, S.I.L.-G., M.F.M. and F.M.G.M.; methodology, S.I.L.-G., F.M.G.M., M.F.M., L.C., V.M.J.-A., E.F. and Á.S.; software, S.I.L.-G., F.M.G.M., Á.S., V.M.J.-A. and L.C.; validation, S.I.L.-G., F.M.G.M., L.C., V.M.J.-A., V.E.R.C. and J.Á.C.M.; formal analysis, S.I.L.-G., M.F.M., F.M.G.M., Á.S., N.T., V.M.J.-A., V.E.R.C. and J.Á.C.M.; investigation, S.I.L.-G., M.F.M., F.M.G.M., L.C., V.M.J.-A., Á.S. and M.M.; resources, S.I.L.-G., F.M.G.M., L.C., E.F. and M.F.M.; data curation, F.M.G.M., V.M.J.-A., Á.S., M.M. and N.T.; writing—original draft preparation, S.I.L.-G., F.M.G.M. and Á.S.; writing—review and editing, S.I.L.-G., F.M.G.M., V.M.J.-A., Á.S., N.T., V.E.R.C., J.Á.C.M., E.F., M.M., L.C. and M.F.M.; visualisation, S.I.L.-G. and F.M.G.M.; supervision, F.M.G.M.; project administration, F.M.G.M.; funding acquisition, F.M.G.M. and M.F.M. All authors have read and agreed to the published version of the manuscript.

Funding: This research received no external funding.

Institutional Review Board Statement: Not available.

Informed Consent Statement: Not available.

Data Availability Statement: The original contributions presented in this study are included in the article. Further inquiries can be directed to the corresponding authors.

Acknowledgments: The authors would like to thank the Programa de Doctorado en Energía Solar of the Universidad de Antofagasta, Chile; the support of ANID-Chile through the research project FONDECYT Iniciación 11230550; the Project DIUDA N° 22430 carried out by the Facultad de Ingeniería of the Universidad de Atacama; and ANID/FONDAP 1523A0006 Solar Energy Research Center SERC-Chile, 2nd extension. Additionally, Victor M. Jiménez-Arévalo appreciates the financial assistance provided by the Vicerrectoria de Postgrado from the Universidad de Santiago de Chile and postdoctorate project DICYT N° 022442PC, Vicerrectoría de Investigación, Innovación y Creación.

Conflicts of Interest: The authors declare no conflicts of interest.

References

1. Holechek, J.L.; Geli, H.M.E.; Sawalhah, M.N.; Valdez, R. A Global Assessment: Can Renewable Energy Replace Fossil Fuels by 2050? *Sustainability* **2022**, *14*, 4792. [CrossRef]
2. Madurai Elavarasan, R.; Pugazhendhi, R.; Irfan, M.; Mihet-Popa, L.; Khan, I.A.; Campana, P.E. State-of-the-Art Sustainable Approaches for Deeper Decarbonization in Europe—An Endowment to Climate Neutral Vision. *Renew. Sustain. Energy Rev.* **2022**, *159*, 112204. [CrossRef]
3. Yue, M.; Lambert, H.; Pahon, E.; Roche, R.; Jemei, S.; Hissel, D. Hydrogen Energy Systems: A Critical Review of Technologies, Applications, Trends and Challenges. *Renew. Sustain. Energy Rev.* **2021**, *146*, 111180. [CrossRef]

4. Hassan, Q.; Tabar, V.S.; Sameen, A.Z.; Salman, H.M.; Jaszczur, M. A Review of Green Hydrogen Production Based on Solar Energy; Techniques and Methods. *Energy Harvest. Syst.* **2024**, *11*, 20220134. [CrossRef]
5. Okolie, J.A.; Patra, B.R.; Mukherjee, A.; Nanda, S.; Dalai, A.K.; Kozinski, J.A. Futuristic Applications of Hydrogen in Energy, Biorefining, Aerospace, Pharmaceuticals and Metallurgy. *Int. J. Hydrog. Energy* **2021**, *46*, 8885–8905. [CrossRef]
6. Dutta, S. Review on Solar Hydrogen: Its Prospects and Limitations. *Energy Fuels* **2021**, *35*, 11613–11639. [CrossRef]
7. Leiva-Guajardo, S.I.; Toro, N.; Fuentealba, E.; Morel, M.J.; Soliz, Á.; Portillo, C.; Galleguillos Madrid, F.M. Contribution of Copper Slag to Water Treatment and Hydrogen Production by Photocatalytic Mechanisms in Aqueous Solutions: A Mini Review. *Materials* **2024**, *17*, 5434. [CrossRef]
8. Sánchez de Rojas, M.I.; Rivera, J.; Frías, M.; Marín, F. Use of Recycled Copper Slag for Blended Cements. *J. Chem. Technol. Biotechnol.* **2008**, *83*, 209–217. [CrossRef]
9. Murari, K.; Siddique, R.; Jain, K.K. Use of Waste Copper Slag, a Sustainable Material. *J. Mater. Cycles Waste Manag.* **2015**, *17*, 13–26. [CrossRef]
10. Shi, C.; Meyer, C.; Behnood, A. Utilization of Copper Slag in Cement and Concrete. *Resour. Conserv. Recycl.* **2008**, *52*, 1115–1120. [CrossRef]
11. Bhandari, R.; Trudewind, C.A.; Zapp, P. Life Cycle Assessment of Hydrogen Production via Electrolysis—A Review. *J. Clean. Prod.* **2014**, *85*, 151–163. [CrossRef]
12. González, C.; Parra, R.; Klenovcanova, A.; Imris, I.; Sánchez, M. Reduction of Chilean Copper Slags: A Case of Waste Management Project. *Scand. J. Metall.* **2005**, *34*, 143–149. [CrossRef]
13. Kim, B.S.; Jo, S.K.; Shin, D.; Lee, J.C.; Jeong, S.B. A Physico-Chemical Separation Process for Upgrading Iron from Waste Copper Slag. *Int. J. Miner. Process.* **2013**, *124*, 124–127. [CrossRef]
14. Gorai, B.; Jana, R.K. Premchand Characteristics and Utilisation of Copper Slag—A Review. *Resour. Conserv. Recycl.* **2003**, *39*, 299–313. [CrossRef]
15. Zhai, Q.; Liu, R.; Wang, C.; Wen, X.; Li, X.; Sun, W. A Novel Scheme for the Utilization of Cu Slag Flotation Tailings in Preparing Internal Electrolysis Materials to Degrade Printing and Dyeing Wastewater. *J. Hazard. Mater.* **2022**, *424*, 127537. [CrossRef]
16. Jena, S.S.; Tripathy, S.K.; Mandre, N.R.; Venugopal, R.; Farrokhpay, S. Sustainable Use of Copper Resources: Beneficiation of Low-Grade Copper Ores. *Minerals* **2022**, *12*, 545. [CrossRef]
17. Gabasiane, T.S.; Danha, G.; Mamvura, T.A.; Mashifana, T.; Dzinomwa, G. Characterization of Copper Slag for Beneficiation of Iron and Copper. *Heliyon* **2021**, *7*, e06757. [CrossRef]
18. Phiri, T.C.; Singh, P.; Nikoloski, A.N. The Potential for Copper Slag Waste as a Resource for a Circular Economy: A Review—Part II. *Miner. Eng.* **2021**, *172*, 107150. [CrossRef]
19. Montoya-Bautista, C.V.; Avella, E.; Ramírez-Zamora, R.M.; Schouwenaars, R. Metallurgical Wastes Employed as Catalysts and Photocatalysts for Water Treatment: A Review. *Sustainability* **2019**, *11*, 2470. [CrossRef]
20. Montoya-Bautista, C.V.; Acevedo-Peña, P.; Zanella, R.; Ramírez-Zamora, R.-M. Characterization and Evaluation of Copper Slag as a Bifunctional Photocatalyst for Alcohols Degradation and Hydrogen Production. *Top. Catal.* **2021**, *64*, 131–141. [CrossRef]
21. Dresp, S.; Dionigi, F.; Klingenhof, M.; Strasser, P. Direct Electrolytic Splitting of Seawater: Opportunities and Challenges. *ACS Energy Lett.* **2019**, *4*, 933–942. [CrossRef]
22. Gao, F.Y.; Yu, P.C.; Gao, M.R. Seawater Electrolysis Technologies for Green Hydrogen Production: Challenges and Opportunities. *Curr. Opin. Chem. Eng.* **2022**, *36*, 100827. [CrossRef]
23. Bamba, J.N.; Dumlao, A.T.; Lazaro, R.M.; Matienzo, D.D.; Ocon, J. Green Hydrogen from Seawater Electrolysis: Recent Developments and Future Perspectives. *Curr. Opin. Electrochem.* **2024**, *48*, 101592. [CrossRef]
24. Montoya-Bautista, C.V.; Alcántar-Vázquez, B.C.; Solís-López, M.; Tabla-Vázquez, C.G.; Morales-Pérez, A.A.; Schouwenaars, R.; Ramírez-Zamora, R.M. Photocatalytic H_2 Production and Carbon Dioxide Capture Using Metallurgical Slag and Slag-Derived Materials. In *Handbook of Ecomaterials*; Springer International Publishing: Berlin/Heidelberg, Germany, 2018; pp. 1–19.
25. Yu, J.; Zhao, C.X.; Liu, J.N.; Li, B.Q.; Tang, C.; Zhang, Q. Seawater-Based Electrolyte for Zinc–Air Batteries. *Green. Chem. Eng.* **2020**, *1*, 117–123. [CrossRef]
26. Pavez, O.; Nazer, A.; Rivera, O.; Salinas, M.; Araya, B. Copper Slag from Different Dumps in the Atacama Region Used in Mortars as Partial Replacement of Cement. *Rev. Mater.* **2019**, *24*, e12349. [CrossRef]
27. Soni, I.; Kumar, P.; Sharma, S.; Kudur Jayaprakash, G. A Short Review on Electrochemical Sensing of Commercial Dyes in Real Samples Using Carbon Paste Electrodes. *Electrochem* **2021**, *2*, 274–294. [CrossRef]
28. Giggenbach, W.F. The isotopic composition of waters from the El Tatio geothermal field, Northern Chile. *Geochemica et Cosmochimica Acta* **1978**, *42*, 979–998. [CrossRef]
29. Munoz-Saez, C.; Manga, M.; Hurwitz, S.; Rudolph, M.L.; Namiki, A.; Wang, C.Y. Dynamics within Geyser Conduits, and Sensitivity to Environmental Perturbations: Insights from a Periodic Geyser in the El Tatio Geyser Field, Atacama Desert, Chile. *J. Volcanol. Geotherm. Res.* **2015**, *292*, 41–55. [CrossRef]

30. Zhao, Y.; Hu, X.; Shi, C.; Zhang, Z.; Zhu, D. A Review on Seawater Sea-Sand Concrete: Mixture Proportion, Hydration, Microstructure and Properties. *Constr. Build. Mater.* **2021**, *295*, 123602. [CrossRef]
31. Mangi, S.A.; Makhija, A.; Saleem Raza, M.; Khahro, S.H.; Ashfaque; Jhatial, A. A Comprehensive Review on Effects of Seawater on Engineering Properties of Concrete. *Silicon* **2021**, *13*, 4519–4526. [CrossRef]
32. Cáceres, L.; Vargas, T.; Parra, M. Study of the Variational Patterns for Corrosion Kinetics of Carbon Steel as a Function of Dissolved Oxygen and NaCl Concentration. *Electrochim. Acta* **2009**, *54*, 7435–7443. [CrossRef]
33. Allibai Mohanan, V.M.; Kacheri Kunnummal, A.; Biju, V.M.N. Electrochemical Sensing of Hydroxylamine Using a Wax Impregnated Graphite Electrode Modified with a Nanocomposite Consisting of Ferric Oxide and Copper Hexacyanoferrate. *Microchim. Acta* **2016**, *183*, 2013–2021. [CrossRef]
34. Li, Z.; Ma, G.; Zhang, X.; Li, J. Characteristics and Chemical Speciation of Waste Copper Slag. *Environ. Sci. Pollut. Res.* **2021**, *28*, 20012–20022. [CrossRef] [PubMed]
35. Tian, H.; Guo, Z.; Pan, J.; Zhu, D.; Yang, C.; Xue, Y.; Li, S.; Wang, D. Comprehensive Review on Metallurgical Recycling and Cleaning of Copper Slag. *Resour. Conserv. Recycl.* **2021**, *168*, 105366. [CrossRef]
36. Giddaerappa; Naseem, K.; Sharath, K.; Hojamberdiev, M.; Sannegowda, L.K. Substrate-Driven Electrocatalysis of Natural and Earth-Abundant Pyrite Towards Oxygen Evolution Reaction. *Electrochim. Acta* **2024**, *475*, 143575. [CrossRef]
37. Gao, M.R.; Zheng, Y.R.; Jiang, J.; Yu, S.H. Pyrite-Type Nanomaterials for Advanced Electrocatalysis. *Acc. Chem. Res.* **2017**, *50*, 2194–2204. [CrossRef]
38. Dai, W.; Pan, Y.; Ren, K.; Zhu, Y.A.; Lu, T. Heteroatom Ni Alloyed Pyrite-Phase FeS2 as a Pre-Catalyst for Enhanced Oxygen Evolution Reaction. *Electrochim. Acta* **2020**, *355*, 136821. [CrossRef]
39. Chang, C.C.; Li, S.R.; Chou, H.L.; Lee, Y.C.; Patil, S.; Lin, Y.S.; Chang, C.C.; Chang, Y.J.; Wang, D.Y. Photoactive Earth-Abundant Iron Pyrite Catalysts for Electrocatalytic Nitrogen Reduction Reaction. *Small* **2019**, *15*, 1904723. [CrossRef]
40. Shit, S.; Samanta, P.; Bolar, S.; Murmu, N.C.; Khanra, P.; Kuila, T. Synthesis of Iron Pyrite with Efficient Bifunctional Electrocatalytic Activity towards Overall Water Splitting in Alkaline Medium. *Bull. Mater. Sci.* **2021**, *44*, 169. [CrossRef]
41. Faber, M.S.; Dziedzic, R.; Lukowski, M.A.; Kaiser, N.S.; Ding, Q.; Jin, S. High-Performance Electrocatalysis Using Metallic Cobalt Pyrite (CoS 2) Micro- and Nanostructures. *J. Am. Chem. Soc.* **2014**, *136*, 10053–10061. [CrossRef]
42. Liu, W.; Hu, E.; Jiang, H.; Xiang, Y.; Weng, Z.; Li, M.; Fan, Q.; Yu, X.; Altman, E.I.; Wang, H. A Highly Active and Stable Hydrogen Evolution Catalyst Based on Pyrite-Structured Cobalt Phosphosulfide. *Nat. Commun.* **2016**, *7*, 10771. [CrossRef] [PubMed]
43. Guo, D.; Wang, Y.; Zhang, Y.; Duan, J.; Guan, F.; Hou, B. Effects of Marine Eutrophication Environment on Microbial Corrosion: A Review. *Mar. Pollut. Bull.* **2024**, *205*, 116637. [CrossRef] [PubMed]
44. Al-Mamun, A.; Baawain, M.S.; Dhar, B.R.; Kim, I.S. Improved Recovery of Bioenergy and Osmotic Water in an Osmotic Microbial Fuel Cell Using Micro-Diffuser Assisted Marine Aerobic Biofilm on Cathode. *Biochem. Eng. J.* **2017**, *128*, 235–242. [CrossRef]
45. Erable, B.; Lacroix, R.; Etcheverry, L.; Féron, D.; Delia, M.L.; Bergel, A. Marine Floating Microbial Fuel Cell Involving Aerobic Biofilm on Stainless Steel Cathodes. *Bioresour. Technol.* **2013**, *142*, 510–516. [CrossRef]
46. Zhang, H.M.; Li, J. Strategies for Overcoming Seawater Adverse Effects on Cathodic Hydrogen Evolution Reaction Electrocatalysts. *Fuel* **2024**, *367*, 131505. [CrossRef]
47. Chen, R.; Yang, C.; Cai, W.; Wang, H.Y.; Miao, J.; Zhang, L.; Chen, S.; Liu, B. Use of Platinum as the Counter Electrode to Study the Activity of Nonprecious Metal Catalysts for the Hydrogen Evolution Reaction. *ACS Energy Lett.* **2017**, *2*, 1070–1075. [CrossRef]
48. Cheng, X.; Li, Y.; Zheng, L.; Yan, Y.; Zhang, Y.; Chen, G.; Sun, S.; Zhang, J. Highly Active, Stable Oxidized Platinum Clusters as Electrocatalysts for the Hydrogen Evolution Reaction. *Energy Environ. Sci.* **2017**, *10*, 2450–2458. [CrossRef]
49. Wang, W.; Wang, W.; Xu, Y.; Ren, X.; Liu, X.; Li, Z. Synthesis of Ni3S4/NiS2/FeS2 Nanoparticles for Hydrogen and Oxygen Evolution Reaction. *Appl. Surf. Sci.* **2021**, *560*, 149985. [CrossRef]
50. She, Z.W.; Kibsgaard, J.; Dickens, C.F.; Chorkendorff, I.; Nørskov, J.K.; Jaramillo, T.F. Combining Theory and Experiment in Electrocatalysis: Insights into Materials Design. *Science* **2017**, *355*, eaad4998.
51. Ma, Z.; Lian, X.; Yuan, K.; Sun, S.; Gu, C.; Zhang, J.L.; Lyu, J.; Zhong, J.Q.; Liu, L.; Li, H.; et al. Pressure-Dependent Band-Bending in ZnO: A near-Ambient-Pressure X-ray Photoelectron Spectroscopy Study. *J. Energy Chem.* **2021**, *60*, 25–31. [CrossRef]
52. Wang, H.; Lee, H.W.; Deng, Y.; Lu, Z.; Hsu, P.C.; Liu, Y.; Lin, D.; Cui, Y. Bifunctional Non-Noble Metal Oxide Nanoparticle Electrocatalysts through Lithium-Induced Conversion for Overall Water Splitting. *Nat. Commun.* **2015**, *6*, 7261. [CrossRef] [PubMed]
53. Skúlason, E.; Karlberg, G.S.; Rossmeisl, J.; Bligaard, T.; Greeley, J.; Jónsson, H.; Nørskov, J.K. Density Functional Theory Calculations for the Hydrogen Evolution Reaction in an Electrochemical Double Layer on the Pt(111) Electrode. *Phys. Chem. Chem. Phys.* **2007**, *9*, 3241–3250. [CrossRef] [PubMed]
54. Jin, Q.; Chen, L. A Review of the Influence of Copper Slag on the Properties of Cement-Based Materials. *Materials* **2022**, *15*, 8594. [CrossRef]
55. Sarfo, P.; Wyss, G.; Ma, G.; Das, A.; Young, C. Carbothermal Reduction of Copper Smelter Slag for Recycling into Pig Iron and Glass. *Miner. Eng.* **2017**, *107*, 8–19. [CrossRef]

56. Das, S.; Hendry, M.J. Application of Raman Spectroscopy to Identify Iron Minerals Commonly Found in Mine Wastes. *Chem. Geol.* **2011**, *290*, 101–108. [CrossRef]
57. Kuebler, K.E.; Jolliff, B.L.; Wang, A.; Haskin, L.A. Extracting Olivine (Fo-Fa) Compositions from Raman Spectral Peak Positions. *Geochim. Cosmochim. Acta* **2006**, *70*, 6201–6222. [CrossRef]
58. Marshall, C.P.; Dufresne, W.J.B.; Rufledt, C.J. Polarized Raman Spectra of Hematite and Assignment of External Modes. *J. Raman Spectrosc.* **2020**, *51*, 1522–1529. [CrossRef]
59. Dudek, M.; Grabarczyk, J.; Jakubowski, T.; Zaręba, P.; Karczemska, A. Raman Spectroscopy Investigations of Ribbeck Meteorite. *Materials* **2024**, *17*, 5105. [CrossRef]
60. Wang, H.; Fang, P.; Wang, Y.; Xin, Y.; Xiong, S.; Liu, S.; Xue, Y.; Zhang, L.; Wan, X. Rapid Determination of Meteorolite Composition Based on X-Ray Phase Contrast Imaging-Assisted Raman Spectroscopy. *Chemosensors* **2023**, *11*, 563. [CrossRef]
61. Santos, D.M.F.; Sequeira, C.A.C.; Figueiredo, J.L. Hydrogen Production by Alkaline Water Electrolysis. *Química Nova* **2013**, *36*, 1176–1193. [CrossRef]
62. Cui, T.; Chi, J.; Liu, K.; Zhu, J.; Guo, L.; Mao, H.; Liu, X.; Lai, J.; Guo, H.; Wang, L. Manipulating the Electron Redistribution of Fe_3O_4 for Anion Exchange Membrane Based Alkaline Seawater Electrolysis. *Appl. Catal. B Environ. Energy* **2024**, *357*, 124269. [CrossRef]
63. Grimm, E. The Study of Copper Corrosion Mechanisms Using Electrochemical Experimental Techniques. Master's Thesis, University of Louisville, Louisville, KY, USA, 2021.
64. Kaiser, S.; Salim, M. A comparative study of chemical and physical properties of copper and copper alloys affected by acidic, alkaline and saline environments. *J. Electrochem. Sci. Eng.* **2020**, *10*, 373–384. [CrossRef]
65. Yaro, A.; Khadom, A.; Idan, M. Electrochemical Approaches of Evaluating Galvanic Corrosion Kinetics of Copper Alloy—Steel Alloy Couple in Inhibited Cooling Water System. *J. Mater. Environ. Sci.* **2015**, *6*, 1101–1104.
66. Khadom, A.; Abod, B.M. Mathematical model for galvanic corrosion of steel-cooper couple in petroleum waste water in presence of friendly corrosion inhibitor. *J. Appl. Res. Technol.* **2017**, *15*, 14–20. [CrossRef]
67. Metikoš-Huković, M.; Babić, R.; Škugor Rončević, I.; Grubač, Z. Corrosion Science Section Corrosion Behavior of the Filmed Copper Surface in Saline Water Under Static and Jet Impingement Conditions. *J. Sci. Eng.* **2012**, *68*. [CrossRef]

www.ingramcontent.com/pod-product-compliance
Lightning Source LLC
Chambersburg PA
CBHW062041220926
51850CB00010B/2629

MDPI AG
Grosspeteranlage 5
4052 Basel
Switzerland
Tel.: +41 61 683 77 34

Materials Editorial Office
E-mail: materials@mdpi.com
www.mdpi.com/journal/materials